ALCOHOL AND DRUG PROBLEMS AT WORK

" . . . one of the most attractive and informative, practical and stimulating publications of comparable size currently available . . . important, ever topical content and excellent presentation."

—*The RoSPA Occupational Safety and Health Journal, July 2003*

ALCOHOL AND DRUG PROBLEMS AT WORK

THE SHIFT TO PREVENTION

Prepared in collaboration with the United Nations Office for Drug Control and Crime Prevention

INTERNATIONAL LABOUR OFFICE · GENEVA

Copyright © International Labour Organization 2003
First published 2003
Second impression 2004

Publications of the International Labour Office enjoy copyright under Protocol 2 of the Universal Copyright Convention. Nevertheless, short excerpts from them may be reproduced without authorization, on condition that the source is indicated. For rights of reproduction or translation, application should be made to the Publications Bureau (Rights and Permissions), International Labour Office, CH-1211 Geneva 22, Switzerland. The International Labour Office welcomes such applications.

Libraries, institutions and other users registered in the United Kingdom with the Copyright Licensing Agency, 90 Tottenham Court Road, London W1P 4LP [Fax: (+ 44) (0) 207 631 5500; email: cla@cla.co.uk], in the United States with the Copyright Clearance Center, 222 Rosewood Drive, Danvers, MA 01923 [Fax (+1) (978) 750 4470; email: info@copyright.com], or in other countries with associated Reproduction Rights Organizations, may make photocopies in accordance with the licences issued to them for this purpose.

Alcohol and drug problems at work: The shift to prevention
Geneva, International Labour Office, 2003

Guide: drug abuse, alcoholism, occupational health, personnel policy.
13.04.7
ISBN 92-2-113373-7

ILO Cataloguing in Publication Data

The designations employed in ILO publications, which are in conformity with United Nations practice, and the presentation of material therein do not imply the expression of any opinion whatsoever on the part of the International Labour Office concerning the legal status of any country, area or territory or of its authorities, or concerning the delimitation of its frontiers.

The responsibility for opinions expressed in signed articles, studies and other contributions rests solely with their authors, and publication does not constitute an endorsement by the International Labour Office of the opinions expressed in them.

Reference to names of firms and commercial products and processes does not imply their endorsement by the International Labour Office, and any failure to mention a particular firm, commercial product or process is not a sign of disapproval.

ILO publications can be obtained through major booksellers or ILO local offices in many countries, or direct from ILO Publications, International Labour Office, CH-1211 Geneva 22, Switzerland. Catalogues or lists of new publications are available free of charge from the above address or by email: pubvente@ilo.org.

Visit our website: http:/www.ilo.org/publns

Typeset by Magheross Graphics, Switzerland & Ireland
Printed and bound in Great Britain by Biddles Ltd.

PREFACE

The many concerns surrounding drug and alcohol problems in the workplace, estimated to cost the economy billions of dollars each year, come together around the broad issues of workers' health, welfare and safety, workplace productivity and legal liabilities. For this reason, the workplace is regarded as an appropriate setting for the formulation and implementation of alcohol and drug policies and programmes.

In recognition of these concerns, in 1987 the 73rd Session of the International Labour Conference adopted a resolution reaffirming the role of the social partners in addressing workplace drug and alcohol problems. Since then, the ILO has been engaged extensively in this field, promoting policy formulation, the improvement of working conditions, awareness creation, supervisory training, and counselling and assistance programmes.

The ILO has been conscious that approaches and solutions must consider the particular circumstances of each situation, especially the different cultural, social and economic factors. Such initiatives should be linked to other efforts designed to improve the working environment. Thus, programmes need to be developed within the framework of corporate culture, health promotion, safety, welfare and productivity. Above all, such programmes should be in harmony with national policy and linked to community action and services.

With its tripartite composition, the ILO is sensitive to the concerns of its constituents on problems related to the workplace. Governments wish to reduce the negative impact on economic development, social costs and public safety. Employers are concerned with safety, productivity and competitiveness. Trade unions wish to safeguard the health, welfare and job security of their members. These interests and concerns converge to form a powerful framework of health, safety, welfare and productivity to support workplace prevention and assistance programmes.

The effectiveness of prevention and assistance programmes is enhanced when they are conducted within an appropriate policy framework which establishes the parameters and the underpinnings for action. Such a framework is more acceptable and easier to implement if it is formulated through joint labour–management consultation and agreement. It was against this background that the Governing Body of the ILO decided to convene a tripartite meeting of experts in January 1995 to adopt a code of practice, *The management of alcohol- and drug-related issues in the workplace*.

The code constitutes a unique document in the domain of demand reduction,[1] backed by a tripartite international consensus. It provides a broad range of practical recommendations and guidance on developing prevention and assistance programmes in the workplace. Every enterprise, whatever its resources, can respond in some way, by selecting appropriate action from the very simple to the highly complex. Decisions depend on fitting the response to the needs, to the resources available within the community and the company, the legislative requirements and cultural factors.

The workplace offers a unique opportunity to reach the most vital and productive segment of the population. Every effort should be made to provide counselling, treatment and return-to-work assistance to those in need. More importantly, workplace programmes should be directed at the entire workforce with the aim of keeping healthy workers healthy. For this simple but essential reason, the ILO has pursued and promoted programmes which have made a paradigm shift to primary prevention.

This programming shift has been successfully tested in a number of countries. Drug and alcohol problems have been recognized as ongoing concerns of management which necessitate sustained action. Joint assessment of the problem and formulation of policy and programmes have become an essential first step. Special focus has been given to awareness creation through a range of information, education and self-assessment initiatives. The shift to prevention has also necessitated a shift in resources. More importantly it has been accompanied by sustained efforts to integrate drug and alcohol prevention into larger programmes such as human resources development, occupational health and safety, workers' education and workers' family welfare.

This book distils the lessons learned from the wide range of projects, implemented in all regions of the world, which promote the paradigm shift to prevention. It provides a step-by-step guide for policy-makers and programme planners in large and medium-sized enterprises. It also sets the stage for further expansion of the prevention approach by aligning action aimed at

---

[1] The reduction of demand for drugs and alcohol, rather than the restriction of their supply.

addressing other psychosocial problems that have an impact on health and safety. Close collaboration and linkage with community services are also encouraged in order to provide support to workers in need of help.

Behrouz Shahandeh
Senior Adviser on Drugs and Alcohol
Programme on Safety, Health and the Environment, SafeWork

CONTENTS

Preface		v
Acknowledgements		xiii
Glossary		xv
1	**Introduction**	1
1.1	Background	1
1.2	The role of the ILO	2
	ILO code of practice	3
1.3	Establishing a workplace substance abuse prevention programme: Key insights	5
2	**Substances of abuse**	9
2.1	Groups of substances of abuse	9
	Alcohol	10
	Drugs	10
	Medications and inhalants	11
2.2	Intoxication, regular use and dependence	12
2.3	Individual physiological effects	13
2.4	Socio-demographic factors	14
3	**Substance abuse and the workplace**	17
3.1	A workplace issue	17
	Risks to the public	19
3.2	Aggravating factors	19
3.3	Addressing substance abuse in the workplace	21

4	The paradigm shift to prevention	23
4.1	Reassessing the problem	23
	Zoning	24
	Self-assessment	25
	Moderate versus heavy drinkers	27
	Health promotion as prevention	28
4.2	Whose programme?	29
	Management	29
	Community and family linkages	29
5	Programme planning	31
5.1	The steering committee	31
5.2	Programme feasibility	33
	External infrastructure	33
	Support within the enterprise	33
	Resources	34
	Needs assessment	34
	Working conditions and workplace attitudes	36
6	Establishing a comprehensive programme	37
6.1	Written policy statement	39
6.2	Training for supervisors, key staff and workers' representatives	41
	Supervisors	42
	Key staff	42
	Workers' representatives	43
6.3	Awareness and education campaigns	43
6.4	Assistance	45
6.5	Evaluation	47
	Process evaluation	47
	Outcome evaluation	47
	Impact evaluation	48
6.6	Alcohol and drug testing	48
7	Sustainability strategies	51
7.1	The importance of sustainability	51
7.2	Alternative strategies	52
	Independent programmes	52
	Integrated programmes	52
	Association of Resource Managers Against Substance Abuse (ARMADA)	53

8	Conclusion	55
8.1	Added benefits	55
8.2	Some success stories	56
	India	56
	Slovenia	57
	Malaysia	57

Annexes

I	Alcohol, alcoholism and alcohol abuse	59
II	Psychoactive substances of abuse	65
III	Signs of substance abuse	73
IV	Self-assessment tools	75
V	Drug testing	79
VI	Websites	89
VII	Print and audio-visual resources	93
VIII	Sample policies on substance abuse	97

Figures

2.1	Individual–environment–drug	14
4.1	The traffic light as metaphor	25
4.2	The paradigm shift from assistance to prevention	26
4.3	Standard drinks	27
5.1	The design, implementation and management of a company prevention programme	32
6.1	Strategies for the prevention of alcohol- and drug-related problems in the workplace	38

Boxes

1.1	Advantages of workplace substance abuse initiatives	2
1.2	ILO code of practice	4
2.1	Major categories of substances of abuse	9
2.2	Problems relating to intoxication, regular use and dependence	12
3.1	Some effects of drugs and alcohol in the workplace	17
4.1	The traffic light analogy	24
6.1	Recommended issues to be covered by a substance abuse prevention policy statement	39
6.2	Topics that could be included in a worker education programme	44
6.3	Assistance for workers with substance abuse problems	46
7.1	Key elements of programme sustainability	51

ACKNOWLEDGEMENTS

This publication, *Alcohol and drug problems at work: The shift to prevention*, has been developed by the International Labour Organization (ILO), with the financial assistance of the United Nations International Drug Control Programme (UNDCP), through the interregional project on Model Programmes of Drug and Alcohol Abuse Prevention among Workers and their Families.

Since the inception of the project, a number of organizations and individuals have provided invaluable guidance and have made contributions which form the foundations of this book. The World Health Organization (WHO) collaborated closely with the project. The international project coordinator and a group of international consultants guided the project and provided expert advice.

At the level of the participating countries, Egypt, Mexico, Namibia, Poland and Sri Lanka, the national teams, made up of a coordinator, a manager and a consultant, supplied key inputs for the adaptation and implementation of the project.

The ILO offices covering the project target countries were continuous sources of policy and administrative support.

The book is based on the project findings, lessons learned and conclusions reached by the evaluation team. A number of experts also contributed to the development of this book by extracting, condensing and presenting ideas.

Deep gratitude is expressed to all for their creativity and contributions.

GLOSSARY

Abuse (alcohol, drug, medication, substance, psychoactive substance): In the DSM-IV,[1] psychoactive abuse is defined as a maladaptive pattern of substance use leading to clinically significant impairment or distress, as manifested by one (or more) of the following over a 12-month period:

- recurrent substance use resulting in a failure to fulfil major role obligations at work, school or home;
- recurrent substance use in situations in which it is physically hazardous;
- recurrent substance-related legal problems;
- continued use despite having persistent or recurrent social or interpersonal problems caused or exacerbated by the effects of the substance.

Alcohol- or drug-related problems: The term can be applied to any range of adverse accompaniments to using alcohol or drugs. "Related" does not necessarily imply causality. The term can be applied to an individual drinker or drug user or to society as a whole. It may be taken to include dependence and abuse, but also covers other problems.

Alcoholism: A primary, chronic health problem with genetic, psychosocial, and environmental factors influencing its development and manifestation. The problem is often progressive and fatal. It is characterized by continuous or periodic impaired control over drinking, preoccupation with the drug alcohol, use of alcohol despite adverse consequences, and distortions in thinking, most notably denial.

Benzodiazepines: Drugs that relieve anxiety or are prescribed as sedatives. They are among the most widely-prescribed medications, including Valium and Librium.

---

[1] *Diagnostic and Statistical Manual of Mental Disorders*, Fourth Edition (DSM-IV), American Psychiatric Association, Washington D.C., 1994.

Chemical dependency: Another term for alcohol or other drug dependency.

Counselling: A process involving a therapeutic relationship between a client who is asking for help and a counsellor or therapist trained to provide that help.

Dependence: The need for repeated doses of alcohol, drugs or medications to feel good or avoid feeling bad. In the DSM-IV, dependence is defined as a cluster of cognitive, behavioural and physiological symptoms indicating that the individual continues use of the substance despite significant related problems.

Drug: In this context the term refers to substances subject to international control, as listed in the various United Nations Conventions and in the Declaration on Drug Demand Reduction (the full lists of scheduled drugs can be downloaded from http://www.incb.org/). It also refers to substances controlled at the national level, as listed in national laws or regulations.

EAP (Employee assistance programme): An employment-based programme that offers assistance to workers, and often their families, with problems that affect or could eventually affect job performance. An EAP can offer assistance and treatment with alcohol- and drug-related problems. It often offers help with other issues that can cause personal distress, such as marital or family difficulties, depression, on-the-job or off-the-job stress, financial problems or legal difficulties.

Early intervention: A therapeutic strategy that combines early detection and the treatment of hazardous or harmful substance use.

Enabling: Any behaviour, direct or indirect, regardless of intention, that allows an individual to continue using alcohol or drugs in a harmful or addictive way.

Harmful use: Patterns of use of alcohol or other drugs for non-medical reasons that result in health consequences and some degree of impairment in social, psychological and occupational functioning.

Illicit or illegal drug: A psychoactive substance listed in the schedules of the international drug control conventions or in national laws and regulations that is of illicit origin.

Occupational health services (OHS): Health services which have an essentially preventive function and are responsible for advising employers, workers and their representatives on requirements for establishing and maintaining a safe and healthy working environment which promotes optimal physical and mental health in relation to work.

Physical dependence (dependency): The state in which the body has adapted to the presence of the drug and withdrawal symptoms occur if use of the drug is stopped abruptly.

Prevention: Intervention designed to change the social and environmental determinants of drug and alcohol abuse by discouraging the onset of drug use and harmful alcohol use, and the progression to more frequent or regular use. Prevention includes:
- creating awareness and informing/educating about substances and their adverse health and social effects;
- assisting individuals and groups to acquire personal and social skills so that they can make informed lifestyle choices;
- promoting supportive environments and healthier, more productive and fulfilling lifestyles.

Psychoactive substance: A substance that, when ingested, can change mood, behaviour and cognition processes. The term and its equivalent, psychotropic drug, are the most neutral and descriptive words for the whole class of substances, licit and illicit. Psychoactive does not necessarily imply dependence producing.

Substance abuse: A maladaptive pattern of substance use leading to clinically-significant impairment or distress such as failure to fulfil major role responsibilities.

Supply reduction: The effort to keep drugs away from people through the work of the police and customs authorities, and by enacting and implementing laws and regulations.

Tolerance: In this context, the capacity of the body to become more or less responsive to alcohol or drugs. Increased tolerance occurs with regular use, requiring the consumption of increasing quantities to produce the effects originally generated by lower quantities. Decreased tolerance occurs when heavy, chronic drinkers experience a sharp drop in tolerance and often become intoxicated after one or two drinks.

INTRODUCTION   1

1.1 Background

The world is witnessing a rising tide of substance abuse. Psychoactive substances are increasingly available. Consumption and drug trafficking are growing. Alcohol and drugs[1] are everywhere. Their abuse is affecting society in ways which were unknown only a few decades ago.

Numerous studies have demonstrated the negative impact of substance abuse on enterprises, as well as on workers and their families. At the worker level, substance abuse can cause impaired health, deterioration in relationships, job loss, and family, legal and financial problems. At the enterprise level, substance abuse has been linked to accidents, absenteeism and lost productivity. Its costs to industry and the community have been estimated in billions of dollars.

Traditionally, policy-makers have emphasized supply reduction as the means of controlling substance abuse, but this has been inadequate. In recent years, there has been a growing focus on reducing demand through programmes designed to keep people away from drugs. Increasingly, the workplace is being recognized as an effective venue for substance abuse prevention activities which influence workers, their families and the community. It is also acknowledged as a good place to address substance abuse problems before they reach the dependence stage requiring medical intervention. Studies have found that people with substance abuse problems are more likely to abandon family and friends than sacrifice their job, which provides the money to pay for alcohol and drugs. The advantages of workplace substance abuse prevention initiatives are set out in box 1.1.

---

[1] Although the term "drugs" may be understood to encompass medications, for the purpose of this publication "drugs" means illegal drugs, which are substances subject to international control by national governments and under the various United Nations Conventions and the Declaration on the Guiding Principles of Drug Demand Reduction.

> **Box 1.1 Advantages of workplace substance abuse initiatives**
>
> - Workplace programmes have the potential to reach the entire working population, from youth to mature adults.
>
> - The workplace mirrors the substance abuse problems of the community. No workplace is immune.
>
> - In the workplace, the target group for the prevention campaign is a captive audience.
>
> - The workplace is an effective location for intervention and for providing support to workers who are experiencing problems of substance abuse. Continued employment is a strong incentive and reliable support for successfully overcoming problems of abuse.
>
> - The greatest potential for reducing alcohol- and drug-related workplace injury exists outside the medical context of hospitals and clinics, because most accidents involve workers who are not yet dependent on alcohol or drugs, and would not, therefore, be in treatment.

1.2 The role of the International Labour Organization

Concern for the welfare of people at work is at the core of the activities of the International Labour Organization (ILO). ILO involvement in combating substance abuse is based on the agreement of its tripartite members – governments, employers' organizations and workers' organizations – that substance abuse has serious consequences for the workplace, and that workplace initiatives are an effective means of preventing and reducing drug and alcohol abuse.

Over the past decade, the ILO has undertaken a series of projects to develop, implement and evaluate a variety of workplace substance abuse models. These efforts have been supported by the United Nations International Drug Control Programme (UNDCP) and donor countries. They have been implemented in different regions of the world, including Africa, Asia, Central and Eastern Europe, the Caribbean and Latin America, covering 40 countries, in cooperation with other international organizations, national governments and non-governmental organizations (NGOs).

One of these projects, Model Programmes of Drug and Alcohol Abuse Prevention among Workers and their Families, piloted a paradigm shift from assistance to prevention. Rather than adhering to the rehabilitative approach, which only targets workers with serious alcohol or drug problems, the programme included the entire workforce and focused on the prevention of abuse.

This publication is based on the lessons learned from that programme's implementation in Egypt, Mexico, Namibia, Poland and Sri Lanka, and is amplified by the principles and recommendations of the ILO code of practice described in detail below, and by the knowledge and experience gained in other ILO programmes of this type. It provides background information and a framework for the development and implementation of a prevention-oriented approach to workplace alcohol and drug problems. Its purpose is to serve as a basic guide for governments and employers' and workers' organizations, as well as employers in large and medium-sized enterprises.

ILO code of practice

The ILO has found that substance abuse prevention programmes in the workplace are more effective when they are developed within a policy framework which clearly defines roles and responsibilities, specifies the scope of activities and explains the kind of assistance available. Experience has shown that these programmes will be easier to implement and more acceptable to workers if they are formulated through joint labour–management consultation and agreement. And, of course, they must be developed within the context of national legislation and country culture.

It was against this background that the ILO developed the code of practice *Management of alcohol- and drug-related issues in the workplace*,[2] published in 1996. The objective of the code is to promote the prevention, reduction and management of problems related to alcohol and other drugs in the workplace; a summary of the key points is given in box 1.2. It is the framework within which the ILO recommends that governments and employers' and workers' organizations develop and implement workplace substance abuse programmes.

The code is a guide to establishing workplace substance abuse prevention programmes. It is not meant to replace or override any more protective international standard or national law or regulation. The implementation of any provision of the code must also take into consideration the particular cultural, legal, social, political and economic circumstances of each country. It must also take into account any collective agreements already in existence.

---

[2] Available from ILO Publications, International Labour Office, CH-1211 Geneva 22, Switzerland, and at http://www.ilo.org/public/english/protection/safework/.

Box 1.2   ILO code of practice

The following are the key points of the code of practice.

- Alcohol and drug policies and programmes should promote prevention, reduction and management of alcohol- and drug-related problems in the workplace. This code applies to all types of public and private employment, including employment in the informal economy. Relevant national legislation and policy should be determined after consultation with the most representative employers' and workers' organizations.

- Alcohol and drug problems should be considered as health problems and therefore dealt with in the same way as any other health problem at work, without discrimination, and covered by the health care systems (public or private) as appropriate.

- Employers and workers and their representatives should jointly assess the effects of alcohol and drug use in the workplace and should cooperate in developing a written policy for the enterprise.

- Employers, in cooperation with workers and their representatives, should do what is reasonably practicable to identify job situations that contribute to alcohol- and drug-related problems, and take appropriate preventive or remedial action.

- The same restrictions or prohibitions with respect to alcohol and drugs should apply to both management personnel and workers, so that there is a clear and unambiguous policy.

- Information, education and training programmes concerning alcohol and drugs should be undertaken to promote safety and health in the workplace and should be integrated where feasible into broad-based health programmes.

- Employers should establish a system to ensure the confidentiality of all information communicated to them concerning alcohol- and drug-related problems. Workers should be informed of exceptions to confidentiality which arise from legal, professional or ethical principles.

- Testing of bodily samples for alcohol and drugs in the context of employment involves moral, ethical and legal issues of fundamental importance, requiring a determination of when it is fair and appropriate to conduct such testing.

- The stability which ensues from holding a job is frequently an important factor in facilitating recovery from alcohol- and drug-related problems. Therefore, the social partners should acknowledge the special role the workplace may play in assisting individuals with such problems.

- Workers who seek treatment and rehabilitation for alcohol- or drug-related problems should not be discriminated against by the employer and should enjoy normal job security and the same opportunities for transfer and advancement as their colleagues.

- It should be recognized that the employer has authority to discipline workers for employment-related misconduct associated with alcohol and drugs. However, counselling, treatment and rehabilitation should be preferred to disciplinary action. Should a worker fail to cooperate fully with the treatment programme, the employer may take disciplinary action as considered appropriate.

- The employer should adopt the principle of non-discrimination in employment based on previous or current use of alcohol or drugs, in accordance with national laws and regulations.

1.3 Establishing a substance abuse prevention programme: Key insights

The ILO has found that if workplace substance abuse prevention programmes are properly developed and implemented, they are good for both employers and workers. As with other programmes addressing employee health, safety and well-being, they are a "win-win" proposition. Substance abuse prevention programmes make good sense: they not only result in a healthier workforce, they also contribute to improved worker morale, a positive enterprise image in the community and increased enterprise productivity.

The results of the Model Programmes of Drug and Alcohol Abuse Prevention among Workers and their Families and subsequent ILO projects can provide valuable insights for those wishing to start substance abuse prevention programmes. The following are key points which emerged in developing and implementing these projects.

- The cultural, economic and political environments change during the course of the project. Therefore, changes in the enterprise setting or changes in the workplace itself are inevitable. Some of these changes may present new opportunities to enhance and expand the enterprise programme. It is the role of the project's steering committee to consider these changes and determine whether and how the prevention programme should be adapted.

- There is no single programme configuration that meets the needs of all businesses. In addition to the cultural, economic and political conditions mentioned above, there are many factors which serve to distinguish one business from another. For example, the industrial sector, size and geographic location, the organizational structure, the age and sex of the workforce, the degree of substance abuse problems and substances that are being abused, and the working conditions and educational and literacy levels of workers, all have an impact on the way in which the prevention programme is organized and implemented.

- Management must ensure that the programme is tailored to meet the needs of the specific workplace and workforce. While sample programmes and policies are helpful, employers should never adopt a programme without a careful review to make sure it is appropriate to their workplaces.

- There must be a policy framework at the national level which addresses the use and abuse of alcohol and drugs. It is also important that this framework supports the development of policies and programmes at the enterprise level. The lack of such a framework will make it much more difficult for employers and workers to address substance abuse.

- Employers' and workers' organizations need to work with their governments to develop laws and regulations that are supportive of workplace substance abuse prevention programmes. At a minimum, governments and employers' and workers' organizations should make sure that there are no legal or regulatory provisions that hinder the development of these programmes.

- Before developing a prevention programme, employers and workers' representatives need to be familiar with legal and regulatory requirements, if any, concerning workplace substance abuse programmes, and especially those that refer to testing and trafficking in the workplace. Employers and workers' representatives should consider meeting with local law enforcement officials to discuss steps employers should take if trafficking in the workplace is suspected.

- For the most part, alcohol and drugs can be addressed in a common substance abuse prevention programme. However, because drugs are by definition illegal, drug use and trafficking need to be given special consideration when developing enterprise programmes.

- Substance abuse prevention programmes should be management led. The basis of these programmes is performance, and performance is a management issue. The development of a programme requires that management determine how best to structure the programme to operate within its specific workplace.

- Healthcare and other professionals trained in substance abuse prevention can be consulted or invited to participate in steering committee meetings. Their input on awareness, education and counselling would facilitate the development of those aspects of the programme. Experts on other programme components such as security and testing can also be consulted.

- The involvement of workers' representatives in the development and implementation of prevention programmes is vital. Their participation contributes to the credibility and acceptance of the programme by workers. Workers' representatives can also play a significant role in promoting awareness and commitment at the worker level.

  If workers are organized, every effort should be made for their representatives to serve on the steering committee. Some countries require that workers' representatives be consulted. In the absence of elected representatives, informal leaders among the workers should be identified and invited to serve on the steering committee.

- Substance abuse prevention programmes should include outreach to workers' families. Families should be made aware of the programme and invited to participate in awareness, education and counselling activities.

  If workers have substance abuse problems, their families are affected. If a member of a worker's family has a substance abuse problem, it will almost certainly have an impact upon the worker's performance. Substance abuse prevention programmes are much more effective if workers' families are also covered. The effectiveness of counselling is also increased when family members are allowed to participate.

In the process of setting up programmes, policies and procedures, it is easy to lose sight of "why". This was best summed up by this quote from a workers' delegate who participated in the experts' meeting to develop the ILO code of practice:

It is important to remember that, when we talk about the effects of alcohol and drug abuse, we are talking about a tragedy of enormous proportions. (...) The people who suffer from these problems are our friends, our co-workers, our colleagues. (...) Perhaps it is possible that some day, in some place, our recommendations will lead to the establishment of a programme that will save someone's life.

SUBSTANCES OF ABUSE 2

2.1 Groups of substances of abuse

There is increasing worldwide concern about substance use, misuse and abuse. Alcohol and drugs are substances used – and abused – in different parts of the world, in a variety of patterns and for a variety of reasons. Box 2.1 presents the major categories of substances of abuse.

Box 2.1 Major categories of substances of abuse

Alcohol
- Beer, wine, liquor
- Different amounts of pure alcohol in products such as non-prescription medications

Drugs
- Cannabis
- Depressants
- Hallucinogens
- Narcotics
- Stimulants

Medications
- Antidepressants
- Painkillers
- Tranquillizers

For prevention and treatment purposes, it is almost irrelevant to distinguish between alcohol and drug abuse. The psychological and behavioural mechanisms of drug dependency are quite similar to those that occur in alcohol dependency. In addition, most people with some degree of alcohol dependency also abuse drugs, while most people who use drugs often use alcohol when their drugs of choice are not available.

There is no acceptable level of drug use. Acceptable levels of alcohol intake vary considerably from country to country and culture to culture. Per capita consumption of alcohol and drugs in a given society is the strongest predictor of the number and types of problems related to substance abuse that will occur in that society. A doubling of the per capita consumption could indicate a three- or four-fold increase in the number of individuals who are drinking at a level that is dangerous to themselves and others.

Alcohol

Alcohol is alcohol: ethanol, the active ingredient in any alcoholic beverage, is the same whether found in beer, wine or liquor (spirits). Studies have shown that harmful effects of alcohol consumption can be numerous. They include drowsiness, slower reaction time, deterioration of motor performance and coordination skills, loss of concentration and memory, and deterioration in intellectual performance. Long-term use of alcohol can cause cirrhosis. Alcohol poisoning and death can occur if alcohol is consumed in excess, whether in a single instance or over an extended period of time. More information on alcohol can be found in Annex I.

Drugs[1]

Drugs are available in practically every country in the world, in varying dosage, form and purity. Drugs which are illegal in one country may be considered less harmful, or even legal, in another. Some drugs can be illegal except when used in religious or cultural contexts. Some psychoactive substances may be local in production and use, and not regulated at all. Some drugs are produced synthetically, such as LSD (lysergic acid diethylamide, or "acid") and MDMA (3,4-methylenedioxymethamfetamine, or "ecstasy"). In general, drugs can be divided into three broad groups according to their pharmacological effects:

- depressants, such as barbiturates, morphine and heroin;
- hallucinogens, such as cannabis, LSD and MDMA (which has stimulating effects as well);
- stimulants, such as amfetamines, cocaine and crack.

---

[1] In this book we refer to individual drugs by their rINN (recommended International Non-proprietary Name).

Substances of abuse

The effects of drugs on the central nervous system vary enormously, and may range from alertness, restlessness, irritability and anxiety to depression, dizziness, sleeplessness, bizarre and sometimes violent behaviour, and distorted perceptions of depth, time, size and shape of objects and movement. The actual impact of any given drug on an individual depends on several factors, including: the amount taken at one time; past drug-taking experience; the mood and activity of the user; the time and place of intake; the presence of other people; the simultaneous use of other drugs; and the manner of administration. Further information on drugs can be found in Annex II.

Medications and inhalants

Although ILO projects have not addressed medications or inhalants, governments, employers and workers need to be aware of the potential impact of their abuse in the workplace.

Medication abuse means that prescribed drugs are not used according to directions, or that they are used by persons for whom they are not prescribed. The side effects of certain medications include distorted perception, deteriorating motor skills, dizziness, fatigue, blurred vision and loss of muscle control. These side effects are heavily influenced by factors such as dosage, use of other substances, health status, lifestyle and body weight. The following are categories of medications with significant side effects which could have an impact on work performance.

Non-prescription medications

- Antihistamines
- Cold and cough medicines
- Sleep aids

Prescription medications

- Antidepressants
- Antihistamines
- Anti-hypertensives
- Anti-rheumatics
- Benzodiazepines (e.g Librium, Valium)
- Cough syrups containing codeine
- Muscle relaxants

- Painkillers (morphine, codeine)
- Tranquillizers

The abuse of these medications is higher among women than among men. In addition, the possibility of medication abuse becomes greater as workers age and increase their use of medications.

Volatile solvents, which are inhaled, constitute a group of substances that are often ignored. They are cheap and easily available, and their use is particularly prevalent among young people, in both developed and developing countries. Commonly used inhalants include glue, solvents and paints.

2.2 Intoxication, regular use and dependence

Box 2.2 outlines the effects of three levels of substance abuse: intoxication, regular use and dependence.

Box 2.2   Problems relating to intoxication, regular use and dependence

Intoxication

A problem relating to intoxication (for example, a drunk-driving accident, a workplace accident, an alcohol poisoning/drug overdose, a domestic argument) occurs as a consequence of one session of alcohol or drug use. Seventy to eighty per cent of acute incidents such as accidents, violence and crime do not occur among heavy drinkers but among moderate and light drinkers, usually as a result of intoxication. Intoxication-related problems do not always result from heavy consumption in a session. An occasional user may have very low tolerance and get highly intoxicated from small amounts of alcohol or drugs.

Regular use

Problems relating to regular use occur with the constant use of a substance at hazardous levels. The user may not be considered clinically dependent on the substance and may not display advanced signs of dependency such as withdrawal. Regular use problems tend to be of a slow, incremental nature. Over time, the constant presence of alcohol or drugs will cause some damage to the user's body. Relationships and work performance may gradually deteriorate along with the user's health and stamina.

Dependence

When people are heavily dependent on a substance, their difficulties are usually evident to others. In general, the degree of dependency is demonstrated by how

> distressed the user becomes when the substance is no longer available. This distress can range from being slightly uncomfortable to being extremely agitated, as is the case when the user experiences withdrawal. Society does not allow uncontrolled use of drugs or alcohol, so the heavily dependent user is likely to feel tension between the desire for more substance use and the restraints on use, which can lead to considerable stress. This, in turn, can result in unreliability and unpredictability and, consequently, in being judged unfavourably by others.

Changing both group behaviour and environmental factors in the workplace can be more effective in preventing intoxication and regular use problems than singling out individuals for intervention. In some cases, even users who are already dependent may respond to prevention programmes. For instance, many dependent smokers have reported being able to cut down when workplace bans on smoking have been introduced.

2.3 Individual physiological factors

The effects of using alcohol and drugs vary greatly among people. Important independent factors include:

- **Body size:** Other things being equal, the larger the body size, the smaller the effect.

- **Gender:** Women generally have a lower tolerance for alcohol than men of the same body weight, mainly due to the different ratio between body mass and muscle and body fluid and fat, which leads to a higher concentration of alcohol in the blood stream of women. Women also have fewer enzymes which metabolize alcohol in the stomach wall, so that most of the alcohol absorption in women takes place in the small intestine, leading to a rapid increase in the blood alcohol concentration. Thus, a woman drinking the same amount of alcohol as a man of identical weight develops a higher concentration of alcohol in the blood, has a longer contact time between the alcohol consumed and the different organs and, therefore, a higher risk of all types of alcohol-related damage.

- **Genetics:** Absence of or damage to certain genes may increase the likelihood of developing a substance dependency.

- **Prior use:** Individuals who drink frequently will be less affected by a specific amount of alcohol than people who do not drink at all or very

Figure 2 Individual-environment-drug

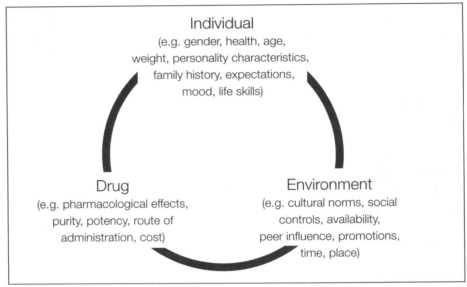

rarely, and who thus have less tolerance. However, individuals who have become highly dependent on a drug tend to use drugs at a much higher rate and level than non-dependent users and, in general, tend to maintain their intake regardless of the restraints a situation may impose upon them.

- **Fatigue:** A person who is not adequately rested has a tendency to be more affected by alcohol consumption than one who is well rested.

2.4 Socio-demographic factors

Substance use and abuse are not evenly distributed throughout any given population. Studies in various countries around the world have shown that alcohol use in the adolescent years is more sporadic throughout the week, reaching peaks at weekends. The consumption pattern of adults between the ages of 35 and 50 is more evenly spread over the weekdays without high peaks. At higher ages, people tend to limit their drinking.

Young people are the predominant users of drugs. In countries such as the United Kingdom and the United States, large numbers of young people (up to 40 per cent of those under 25 years of age) admit to having used drugs at least once. In some European countries, the percentage of young people reporting drug use is lower, although still considerable. In many countries, however, there is no reliable data on drug use.

There is no conclusive evidence of cross-national correlations between income and education and substance abuse. The relationship between exclusion and substance abuse is complex and self-reinforcing: the use and abuse of drugs is often concentrated in areas of social exclusion (because of income level or ethnic group, for example), and drug users themselves usually face exclusion.

There are cultural (social and religious) factors that influence the levels of abuse of different substances. These factors are often gender specific, hence the higher incidence of men reporting substance abuse than women.

SUBSTANCE ABUSE AND THE WORKPLACE

3

Alcohol and drug abuse affects ordinary people. It occurs in the majority of organizations at all levels, from top management to rank and file workers.

3.1 A workplace issue

Alcohol and substance abuse in the workplace can put employees, their co-workers and their enterprise at risk. Depending on the sector and occupation, this risk can sometimes be extended to the general public. Some of the effects on the workplace are given in box 3.1.

Box 3.1   Some effects of substance abuse in the workplace

Accidents

- Up to 40 per cent of accidents at work involve or are related to alcohol use
- Drug-using workers are more likely to be involved in an accident at work than workers who do not use drugs

Absenteeism

Absence from work

- Drug-using or heavy-drinking workers are more likely than other workers to be absent without permission
- Drug-using or heavy-drinking workers are absent from work more often than other workers
- Drug-using or heavy-drinking workers are more likely than other workers to have absences of eight days or more

---

- A large auto manufacturer found that workers who were heavy drinkers/alcoholics had regular patterns of absenteeism on Mondays (error rates were also higher on Mondays)

Tardiness

- Drug-using or heavy-drinking workers are more likely to arrive at work late and leave early than other workers

Strains on co-workers

- Increased workload to compensate for drug-using or heavy-drinking workers
- Higher safety risks due to intoxication, negligence and impaired judgment of drug-using or heavy-drinking workers
- Disputes and grievances
- Lost time leading to decreased productivity
- Intimidation and trafficking in illicit drugs at the worksite
- Violence
- Theft

Replacement costs

- Drug-using or heavy-drinking workers are more likely than other workers to have worked for three or more employers in the past year
- Drug-using or heavy-drinking workers are more likely than other workers to have voluntarily left an employer in the past year
- Drug-using or heavy-drinking workers are more likely than other workers to have been fired by an employer in the past two years

Workers' compensation costs

- Drug-using or heavy-drinking workers are more likely to file a workers' compensation claim than other workers

Output

Both intoxication and post-use impairment ("hangover effect") impact the following functions, which are relevant to work performance.

- Reaction time (reactions are slower)
- Motor performance (clumsy movements and poor coordination)

- Sight (blurred vision)
- Mood (aggression or depression)
- Learning and memory (loss of concentration)
- Intellectual performance (impairment of logical thinking)

Little is known about medication use and its impact on working populations, but the limited research suggests a significant and growing problem with serious effects on workplace safety, health and productivity.

Risks to the public

There are certain occupations and jobs in which the use of alcohol or drugs in the workplace can increase health and safety risks not only to the employee concerned and to his or her co-workers, but also to the public. Examples of employment sectors in which the use of alcohol or drugs may be particularly problematic include the following:

- Transportation (road – including local delivery vehicles – rail, air and marine)
- Law enforcement and security
- Fire service
- Health service
- Hazardous industries (nuclear, chemical or biological research; manufacturing of hazardous substances)

3.2 Aggravating factors

Although the use of alcohol and drugs among workers can have many different origins, there are working conditions that can promote or increase alcohol and substance use. These include:

- Extreme safety risks
- Shift or night work
- Work in remote locations
- Travel away from home

- Changes in tasks or speed of handling machines
- Role conflicts
- Workload (either too much or too little)
- Unequal rewards
- Job stress
- Boredom and lack of creativity, variety or control
- Unsatisfactory communication
- Job insecurity
- Unclear roles

Combinations of these factors are more prominent in some sectors than others, and indeed studies indicate that rates of alcohol or drug use are higher among workers in certain occupations.

- Managers, sales staff, physicians, lawyers, bartenders and entertainers are at high risk of dependency because of social pressure, availability of substances of abuse and lack of supervision.
- For workers who operate powerful equipment, function under extreme conditions or have special responsibility for the safety of others, any substance use could be particularly problematic because of the heightened safety demands and potential impact of an accident on large numbers of people.
- Heavy drinking is more prevalent among men than women, and male-dominated professions such as construction and forestry appear to reinforce this tendency.
- Younger workers are unfamiliar with enterprise cultures and certain working conditions, and are more vulnerable to peer pressure. In addition, many are more susceptible to the effects of alcohol or other substances.

While the effects of certain working conditions can promote or increase substance and alcohol use, the way in which individual workers respond to these conditions varies enormously. Some workers are more vulnerable to the influence of the environment and substances, and have fewer skills for coping with problems than others.

3.3 Addressing substance abuse in the workplace

Some companies consider that disciplinary action, including dismissal, is an adequate response to the problem of substance abuse. However, this punitive approach has a number of serious disadvantages:

- The law, industrial tribunals and workers' organizations increasingly demand constructive responses from employers.
- Dismissal is costly, as it may result in the loss of valued workers.
- Recruiting and training new staff is expensive and time consuming.
- Dismissal is likely to make the problem worse, in that the individual concerned becomes a burden to the community, and the costs of this burden are passed back to the employer as a member of the community.
- If the work environment is contributing to the problem, dismissal will not solve it.

Dismissal may be perceived by other workers as extreme and inappropriate, encouraging them to cover up for co-workers who are drinking heavily or taking drugs.

Prevention programmes address substance abuse by methods that need not be expensive and are likely to produce economic and other benefits. By responding constructively, employers can realize substantial savings by decreasing costs resulting from absenteeism, accidents, production slowdowns, poor work quality, grievances, dismissals, compensation claims and turnover. Recovering dependent workers and former problem drinkers also derive great benefits in terms of improved family life, better relations with co-workers, increased earning capacity, greater job security and restored self-respect.

THE PARADIGM SHIFT TO PREVENTION 4

The basic philosophy of workplace substance abuse prevention programmes is that substance abuse is a preventable health problem. By making prevention a focus in the workplace, fewer workers will develop substance abuse problems.

There are two basic reasons for implementing a substance abuse prevention programme.

- Prevention initiatives emphasize worker health, well-being and safety. This is a positive approach. Both management and labour can support prevention initiatives in a manner which is non-threatening to employers, workers' representatives and workers.

- Introducing a prevention programme benefits the enterprise. Epidemiological studies have demonstrated that the incidence of alcohol- or drug-related problems is correlated to the amount of alcohol consumed; the greater the consumption, the greater the number of problems such as accidents, absenteeism, violence and harassment. In addition, implementing a prevention programme is less costly than the treatment of dependent workers. As a result, enterprise productivity and competitiveness increase.

4.1 Reassessing the problem

The traditional focus of workplace substance abuse programmes has been on providing treatment and rehabilitation to dependent workers. However, most workers are not dependent on alcohol or drugs. In some countries, it is estimated that only ten in every 100 people have an alcohol or drug problem, and, of those ten, only three have become dependent. The paradigm shift away from intervention, directed at dependent workers, to prevention, directed at the whole workforce, was introduced in the ILO's project, Model Programmes of Drug and Alcohol Abuse Prevention among Workers and their Families. The major concepts underlying this project are discussed below.

Zoning

Using the analogy of a traffic light, the level of a worker's use of alcohol and drugs can be divided into one of three zones – the green zone, the amber zone or the red zone – as described in box 4.1.

Moving from the green zone to the red zone is a process that occurs over time. It can take only a few months, especially in the case of certain drugs, or up to several years. The progression from one zone to the next is not marked by a single event or a specific amount of the substance consumed. In moving from green to amber, moderate use gradually develops into problem use. As substance abuse increases, the individual moves through the amber zone and may eventually become dependent. Because there is no uniquely identifiable event, only health professionals trained in substance abuse can determine when an individual becomes dependent.

Traditional workplace substance abuse programmes focus almost exclusively on providing assistance to workers in the red zone. As a result, available resources are concentrated on too few workers, too late.

Box 4.1   The traffic light analogy

Green zone

Workers who don't drink alcohol or drink only in moderation fall within the "green zone". Their alcohol use takes into account the demands of the setting in which consumption takes place or in which they will be immediately after consumption (for example, traffic rules, work demands). These workers can safely continue their drinking patterns – they have a green light.

Amber zone

Workers in the "amber zone" use alcohol to excess or use drugs (illegal drug use has no green zone), but they are not yet dependent on these substances. Because they are not yet dependent, they are generally able to change their habits by themselves or with the assistance of professional counsellors. Workers in this zone need to exercise caution in their use of alcohol and drugs. They need to slow or perhaps stop their use of alcohol, and, in the case of drugs, stop use altogether. These workers have an amber light.

Red zone

Red means stop. Employees in the "red zone" are dependent on one or more substances (alcohol or drugs). They experience serious problems at work and in other areas of their lives. Workers in the red zone usually need professional assistance or treatment to address their dependency so they can return to normal functioning at work and in private.

The paradigm shift to prevention

Figure 4.1 The traffic light as metaphor

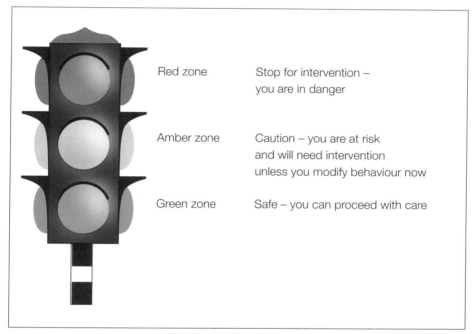

By shifting to a substance abuse *prevention* programme, the focus expands to the entire workforce; the emphasis is on workers in the green and amber zones. Prevention programmes dedicate the majority of available resources to those workers, while still providing help for workers in the red zone (figure 4.2 on page 26).

Self-assessment

Self-assessments can be important tools to help educate workers about their own level of substance use, as well as to help understand the degree of abuse in the workforce. For example, the concept of standard drinks and of counting drinks consumed can be used to provide workers with a simple self-assessment tool to determine the level or "zone" of their drinking.[1] The concept of a standard drink means that alcoholic beverages in their "proper" containers, be it bottle, wine glass or shot glass, all contain the same amount – that is, the

---

[1] There are limitations to using this method as part of a sensible drinking campaign: different national measures of a "unit" and of a "standard" drink exist; a drink poured at home rarely corresponds to the "standard"; there are different alcohol levels in different beverages; and the absorption of units of alcohol is dependent on body weight and gender.

Figure 4.2 The paradigm shift from assistance to prevention

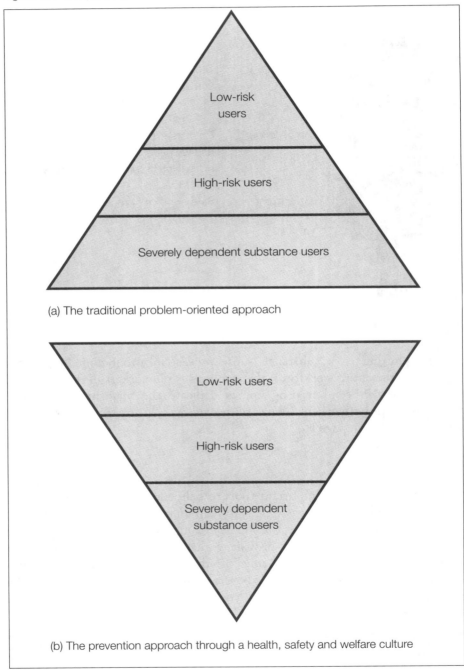

(a) The traditional problem-oriented approach

(b) The prevention approach through a health, safety and welfare culture

Figure 4.3 Standard drinks

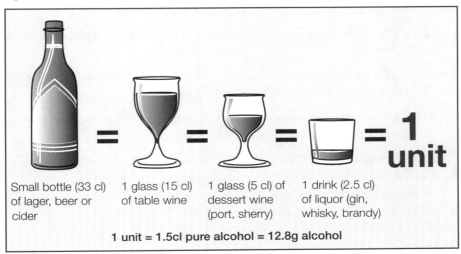

Small bottle (33 cl) of lager, beer or cider = 1 glass (15 cl) of table wine = 1 glass (5 cl) of dessert wine (port, sherry) = 1 drink (2.5 cl) of liquor (gin, whisky, brandy) = 1 unit

1 unit = 1.5cl pure alcohol = 12.8g alcohol

same number of units – of pure alcohol (see figure 4.3). Daily or weekly intake can thus be easily measured. The "amber zone" would be an intake of 22 units or more of alcohol a week for men, and 14 or more units a week for women. More substance abuse self-assessment tools are given in Annex IV.

Moderate versus heavy drinkers

Chronic substance abusers can have a dramatic economic and social impact on the workplace. However, even moderate use of alcohol or drugs can have a negative effect. The correlation between workplace performance and alcohol has been particularly well documented. Laboratory studies clearly indicate that the amount of alcohol found in a single commercial cocktail affects performance and social behaviours that are relevant to workplace performance.

When moderate or even occasional drinkers drink to excess, the impact in the workplace can be just as dramatic as that of chronic substance abusers. And because the total number of workers in the green zone far outnumbers the heavy and dependent drinkers, the *total number* of alcohol-related incidents caused by moderate or occasional drinkers can be greater than the *total number* of incidents caused by heavy or dependent drinkers. Therefore, reaching these workers through a substance abuse programme that focuses on prevention and early intervention will substantially reduce the overall number of incidents.

Furthermore, performance can still be affected when blood alcohol levels return to zero. This was clearly demonstrated in a 1991 study in which airline pilots were asked to perform routine tasks in a flight simulator under three alcohol test conditions, with the following results:

- **First test:** Before any alcohol ingestion, 10 per cent could not perform all tasks correctly.

- **Second test:** After reaching a blood alcohol concentration of 0.1/100ml, 89 per cent could not perform all tests correctly.

- **Third test:** After all alcohol had left their systems 14 hours later, 68 per cent could not perform all tasks correctly.[2]

This "hangover effect" can inhibit the ability to perform at an acceptable level, and it is especially dangerous in high-risk situations as it can jeopardize co-workers and damage equipment. The abuse of alcohol and drugs in some occupations can also put the general public at risk.

Health promotion as prevention

Early intervention programmes which focus on the long-term medical effects of substance abuse have been found to have little impact on people who are already using drugs or abusing alcohol. Health promotion programmes which are aimed at the workforce as a whole can target a greater number of workers whose substance use is not yet problematic, and thus can be used as a powerful form of substance abuse prevention.

The focus of a health promotion programme is the encouragement of a healthy lifestyle, which includes teaching people to use alcohol appropriately and to avoid drug use. Its effects are long-lasting: as workers adopt a new lifestyle and experience better health, they have more incentive to maintain this status. An important part of such programmes is the provision or support of recreational and social activities for participants, either in the workplace or the community.

In many countries, the foremost health problem has become the prevention of the spread of HIV/AIDS. The risk of infection is closely linked to alcohol and drug use. People under the influence of alcohol or drugs are more likely to have unprotected sex. Drug users who inject their drugs are particularly at risk of infection, as sharing needles can transfer the virus from person to person. The relationship between substance abuse and HIV should be stressed throughout the programme.[3]

---

[2] J.G. Modell and J.M. Mountz, "Drinking and flying: The problem of alcohol use by pilots", in *New England Journal of Medicine*, Vol. 323, No. 7, 1990, pp. 455–461.

[3] Additional information on HIV/AIDS policy and programme development, including the ILO code of practice *HIV/AIDS and the world of work*, can be found on the ILO website at http://ilo.org/public/english/protection/trav/aids/.

4.2 Whose programme?

Management

Although connected to health promotion, substance abuse prevention programmes should be the responsibility of management, not the healthcare community. The focus of these programmes is on performance, which is a management issue. When performance is impaired because of substance abuse, it is important for management to take appropriate action.

Management must take the lead in policy development and implementation. A comprehensive workplace programme on substance abuse should be integrated into management strategies and related programmes, such as occupational safety and health, in order to assure long-term sustainability.

The support of the highest levels of management is essential to the success of any workplace substance abuse prevention programme. Management's awareness, commitment and example are the best guarantees for the stability of the programme. Without them, long-term programme implementation and effectiveness will be hampered.

Community and family linkages

There is a direct relationship between the workplace and the community. Recognizing this, enterprises of every size increasingly formulate their roles in terms of corporate responsibility and business ethics. Businesses are more than just producers and sellers of goods and services. Many seek to be active players in their communities, resulting in partnerships which benefit both society and the world of work.

The workplace reflects the level of permissiveness in society regarding the use of alcohol and drugs. Conversely, the degree of tolerance toward substance abuse in the workplace has a strong influence on workers' use of alcohol and drugs in their free time, and hence on the communities in which they live.

In matters of substance abuse, the well-being of the workplace and the family go hand in hand, and enterprises are well aware of the relationship between work and private life. Problems related to alcohol and drugs can arise as a consequence of personal, family or social circumstances, or they can be due to working conditions. Most likely, they result from a combination of these factors. Whatever the causes, substance abuse has an adverse effect on the health and well-being of workers and on the competitiveness of enterprises.

Looking ahead to the needs of tomorrow's workers is a key element of corporate responsibility. People enter the workforce young, when the likelihood that they will experiment with alcohol and drugs is higher than average. Because of their youth, they are more vulnerable to the effects of substances of abuse and to peer influence. An early investment in education is critical to workplace substance abuse prevention. This can be done most effectively if enterprises, schools, healthcare organizations, community organizations and the public sector work together.

Enterprises can support family and community activities on substance abuse prevention in many ways, including:

- school-based activities during and after school hours;
- awareness materials targeting parents and children;
- training and employment initiatives for recovering workers;
- lobbying within the community to establish regulations supportive of substance abuse prevention programmes and related services;
- community recreation facilities and activities.

The treatment and rehabilitation of dependent workers is a very good example of how important it is for enterprises and communities to work together. Appropriate facilities must be available within the community to treat workers who are dependent on alcohol or drugs. At the same time, community-based treatment and rehabilitation services can benefit from their relations with enterprises to assist clients seeking jobs. Employers can provide advice on employment opportunities in the community and training programmes to help recovering workers qualify for available jobs.

PROGRAMME PLANNING 5

The success of a substance abuse prevention programme is directly related to the quality of its planning and preparation. Inadequate planning and preparation will result in problems during implementation and could even jeopardize its credibility.

A programme can be divided into three overlapping phases: design, implementation and management. Figure 5.1 (see page 32) gives a schematic overview of how the initial planning phase is linked to the other stages of the programme, which are dealt with in the chapters to follow.

5.1 The steering committee

Initially, one or two concerned individuals may drive the process of establishing a substance abuse prevention programme. However, to ensure comprehensive planning and acceptance of the programme, a broad-based steering committee should be formed. The role of the steering committee is to define goals and objectives, and to provide overall guidance and supervision for the development and implementation of the programme.

The steering committee should include workers or their representatives. Without their support, it will be more difficult to implement a credible and effective programme. Other members should include representatives of top management and programme areas within the organization who have a special interest in a substance abuse prevention programme. It may also be helpful to have external workplace substance abuse prevention specialists sit on the steering committee.

Each steering committee member should have at least a basic understanding of substance abuse, the principles of prevention and the impact of substance abuse on the workplace. It may be necessary to provide orientation and training to steering committee members so that they have a common understanding and level of knowledge.

Figure 5.1 The design, implementation and management of a company prevention programme

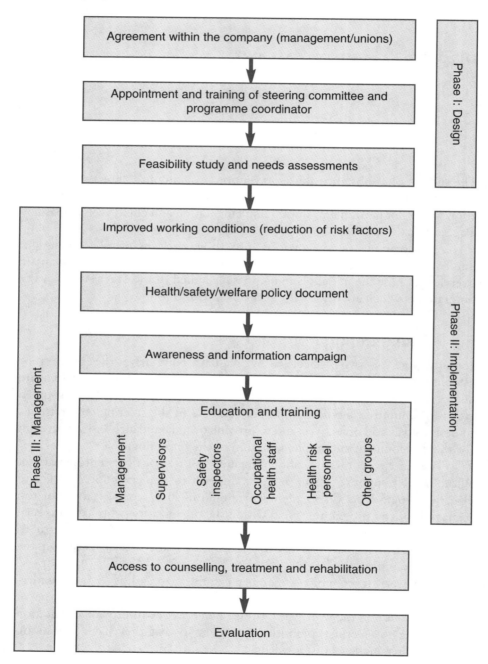

It will be necessary to identify a programme director. The task should be assigned to a staff member, such as an occupational health professional, who has a good understanding of substance abuse and prevention principles.

5.2　Programme feasibility

Before actual planning and preparation begin, a feasibility study needs to be completed or overseen by the steering committee in order to determine whether the necessary external infrastructure, enterprise support and resources exist.

External infrastructure

The presence of an adequate infrastructure outside the organization is essential. Such an infrastructure encompasses appropriate legislative and regulatory provisions as well as adequate community resources to support counselling and treatment of workers with substance abuse problems. Perhaps most importantly, the community itself needs to be engaged in alcohol and drug prevention initiatives so that a workplace prevention programme reflects its values. The business community cannot create such an infrastructure on its own. This requires the cooperation of government, healthcare professionals, workers' representatives and the local community.

Support within the enterprise

Enterprise support refers primarily to the commitment of top management, the cooperation and participation of workers' representatives, and the knowledge and expertise of key staff.

Top management

Without the support of the owner or chief executive officer of the enterprise, it will be difficult, if not impossible, to start or sustain a prevention programme. Top management needs to make decisions on the nature, priority and purpose of a substance abuse programme, the strategy for reaching its goals and the personnel who will be responsible for its development and implementation. Managers and supervisors need to understand that top management is committed to implementing the programme, and that they will be held accountable for fulfilling their responsibilities.

Workers' representatives

Workplace substance abuse prevention programmes should be developed in close cooperation with workers and their representatives. In some countries, the law may require their participation.

Involving workers and their representatives may well be critical to workers' acceptance of the programme. Employers and workers' representatives share the goal of a safe and healthy workplace. Workers' representatives can achieve far more if they involve themselves in the process of restoring workers' health and well-being than by avoiding the issue until jobs are in jeopardy.

Key staff

The design and implementation of a substance abuse prevention programme cannot be the sole responsibility of one manager. A programme's development and success will depend on the involvement and support of staff from human resources, occupational safety and health and the medical unit, and the legal, security and welfare departments.

Resources

Essential resources, such as people, time and money, are necessary for the development and implementation of a prevention programme. For the programme to be effective, the enterprise must make a financial commitment. Studies of traditional substance abuse programmes indicate that the return on investment over time exceeds the cost of programme development and implementation. The return on investment for a prevention programme could be significantly higher, because it covers the entire workforce and is much less costly to operate than a traditional programme.

Needs assessment

The various stakeholders of an enterprise (for example, management, workers' representatives and key personnel) often hold different views on the nature and extent of the situation and the most effective strategy for addressing it (such as a strict policy with emphasis on rules and sanctions, or an education and assistance approach). A thorough assessment of the situation serves as a first step toward the development of a common understanding and, consequently, of a more effective and efficient programme.

A needs assessment analyses an existing situation. The purpose is to collect baseline information, discuss the results and reach a consensus on programme content and strategy. A thorough needs assessment makes more apparent the appropriate nature and content of a workplace substance abuse prevention programme and strategies for its implementation.

A needs assessment may be more or less formal and structured. Both external and internal conditions should be assessed.

External needs assessment

A needs assessment of external conditions could provide information on the following:

- data on the nature and extent of substance abuse at the national and local level;
- the availability of substances of abuse;
- resources available at the national or local level to support a workplace substance abuse prevention programme (such as subject matter experts, training staff, counselling and treatment facilities);
- national laws and regulations that relate to workplace substance abuse;
- substance abuse prevention programmes and activities in other enterprises in the community or within the industry;
- prevailing values and norms on substance use and abuse in the country and the local community.

Internal needs assessment

A substance abuse policy should not be based exclusively on national or local data. Each enterprise has its own culture. Moreover, management and workers will agree more readily with a proposed policy and programme when data are more specific. Examples of enterprise-specific information that can be collected during a needs assessment include:

- availability of alcohol or other substances at the workplace or in the immediate vicinity of the organization;
- existing patterns of use and abuse of alcohol or drugs by employees;
- records of accidents, incidents, absenteeism, sick leave, aggression, violence and terminations;
- medical or other health-related information, including benefits utilization and compensation claims.

The extent to which an enterprise has records containing this information will vary. Other methods for gathering information may include interviews with key persons in the enterprise, or worker and management questionnaires.

It is also useful to review the enterprise's ability to handle substance abuse problems as they arise and the potential impact of such problems on the enterprise. This complements the needs assessment and examines the following:

- risk factors related to the organization of work (for example, working alone and without supervision, shift work, lengthy absences from home and family), the nature of the work (for example, exposure to physical danger, high levels of stress) and the demographics of the workforce (age, education, sex);
- existence, content and awareness of any existing substance abuse programme;
- availability of assistance;
- level of knowledge within the enterprise about signs and symptoms of substance abuse and crisis-management skills and procedures.

Working conditions and workplace attitudes

Assessing the working conditions in an enterprise should be an integral part of the needs assessment. Working conditions can aggravate or mitigate substance abuse, as is highlighted in Chapter 3. Much research has been done on workplace factors that contribute to or reduce the use of alcohol and drugs. Among the most significant factors are the workplace drinking and drug-taking culture and the availability of substances of abuse.

The drinking and drug-taking culture refers to the way drinking or drug-taking behaviour is learned, unlearned or practised in the workplace. Every workplace develops norms and values that guide when and where workers can drink or use drugs, acceptable reasons for drinking, and forms of social control. The impact of the drinking culture can be positive or negative; it can reduce or eliminate workplace substance use or encourage increased and even excessive use.

Assessing workplace attitudes and working conditions is an important part of understanding the potential level of risk of substance abuse among workers, though individually workers may respond to these factors in very different ways.

ESTABLISHING COMPREHENSIVE SUBSTANCE ABUSE PREVENTION PROGRAMMES 6

The feasibility study and needs assessments establish the parameters and scope of a prevention programme. Once this groundwork is complete, a comprehensive programme can be planned that will match the needs and means of the individual enterprise concerned.

A prevention programme plan should include the following elements:

- improvement in working conditions identified as risk factors in substance abuse;
- a written policy statement;
- management and supervisory training;
- awareness and education for workers;
- self-assessments;
- a mechanism through which workers with substance abuse problems can receive assistance in terms of counselling, treatment and rehabilitation;
- evaluation.

A comprehensive programme may also include a testing element, depending on legislative or regulatory provisions and on the nature of the work.

Figure 6.1 gives an example of one form of prevention programme plan which includes the elements listed above. The programme targets all personnel in the primary stages, and then focuses in on different groups within the workforce, according to their needs, using specific tools.

There is an important difference between the respective abilities of small, medium-sized and large enterprises to develop and implement workplace prevention programmes. A large enterprise can generally establish and implement a comprehensive programme using existing communication channels, in-house expertise and available resources.

Figure 6.1 Strategies for the prevention of alcohol- and drug-related problems in the workplace

	Target groups	Problem dimension	Tools	
Primary prevention (public)	All personnel	None: How to cope with alcohol?	Written policy	**Awareness raising and structural/work culture change**
			Attitudes	
	All new employees	Positions with no or low level supervision	Alcohol and drug lifestyle education programmes	
	High-risk categories	Safety risks	Improved working conditions	
		Availability and exposure	Special alcohol and drug education programmes	
Secondary prevention (confidential)	Individuals	Possible post-alcohol use impairment (hangover)	Simple advice	**Assistance**
		Possible health problems	Early identification and assistance	
		Harmful drinking	"Check-ups"	
			Training	
Tertiary prevention (formal)	Identified heavy drug and alcohol users	Safety risks	Treatment programmes	**Rehabilitation**
		Health problems	Employee assistance programmes	
		Instability at work	Family assistance	

Smaller enterprises, however, usually lack the sophistication and resources to implement a comprehensive prevention programme. For example, many small businesses do not have policy statements for any area of operation, so the need for one on substance abuse prevention is not easily understood. Working with small businesses requires adapting the initial programme to those elements with which owners are comfortable, and identifying community resources that can provide the necessary support. For further information, please consult the ILO publication *Workplace substance abuse prevention in small businesses*.[1]

---

[1] Available through ILO SafeWork, International Labour Office, CH-1211 Geneva 22, Switzerland, and at http://www.ilo.org/public/English/protection/safework..

6.1 Written policy statement

A workplace substance abuse prevention policy sets out the objectives and goals of the programme, its structure and elements and the rules and responsibilities in a clear and concise manner understandable to all. A policy statement usually addresses the topics set out in box 6.1.

> Box 6.1 Recommended issues to be covered by a substance abuse prevention policy statement
>
> **Programme rationale**
>
> Reasons for establishing a substance abuse prevention programme vary among enterprises. The rationale may include worker health, safety and well-being; the health and safety of customers; product quality; and recent incidents related to substance abuse. The rationale in turn determines programme goals, objectives, content, strategies and method of evaluation.
>
> **Objectives and goals**
>
> A workplace prevention programme requires clearly defined objectives, and goals which are moderate, realistic, specific and measurable. Goals may measure the process, for example the number of training courses, or the outcome, such as fewer accidents. It is also important to differentiate between short-term goals, such as raising awareness or changing drinking behaviour, and long-term goals, such as reducing absenteeism and improving productivity.
>
> **Coverage**
>
> The programme should cover all employees. It is especially important that all managers, including top management, be included; management should be seen as setting the example.
>
> Substance abuse problems affect the entire family. And as workers change their drinking habits, the relationships among family members will change. Therefore, employers should give serious consideration to including family members in the coverage of their programmes.
>
> **Confidentiality**
>
> The policy should state that personal information on employees taking part in the programme will be treated in a confidential manner.
>
> **Roles and responsibilities**
>
> The implementation of the prevention programme may involve several managers, reflecting the different enterprise units represented on the steering committee.

Supervisors also have significant responsibilities in programme implementation. The roles and responsibilities of all people involved in the development and implementation of the prevention programme should be clearly defined.

Psychoactive substances covered

Depending on the psychoactive substances that are being abused in a specific country or community and the rationale for having a prevention programme, employers need to determine which substances will be covered by the programme (for example, alcohol, specific drugs).

Availability, possession and use

The policy needs to clearly state which substances can be brought onto the enterprise property, which substances can be used during working hours but must be reported to the supervisor or medical staff, and whether or not alcohol is available in enterprise eating facilities, break rooms and at company-sponsored events.

Training, awareness and education

The policy should clearly state the kinds of training, awareness and education activities available to managers, supervisors, workers' representatives and workers.

Assessment, counselling and treatment

The kinds of assistance available to workers who are experiencing alcohol or drug abuse should be identified. Assistance can take the form of counselling and assessment, referral to treatment and rehabilitation facilities, and support for workers who have completed treatment and rehabilitation and are ready to re-enter the workplace.

Testing

The testing of bodily fluids for alcohol or drugs is very controversial, and restrictions or prohibitions on testing may exist in national laws and regulations. On the other hand, laws and regulations may require that workers in certain occupations or jobs be tested for the presence of substances of abuse. The policy statement should clearly state whether testing is to be done, who will be tested, under what conditions and for what psychoactive substances. Because of the complexities involved in establishing and implementing a testing programme, a professional in alcohol and drug testing should be consulted.

Impact on job status

The policy should clearly state that workers who have had previous substance abuse problems or who are successfully undergoing counselling and treatment for a current substance abuse problem will have employment and advancement opportunities equal to those of other workers. There may, however, be national laws and regulations in place that restrict the placement of workers with previous substance abuse problems in certain occupations or jobs.

Violations of the policy

Information, education, counselling, treatment and rehabilitation should be preferred and implemented prior to disciplinary action. However, the employer does have the right to invoke sanctions should violations of the substance abuse policy occur or should the troubled employee choose not to cooperate and the performance problems or misconduct continue. The consequences of policy violations should be specified in the policy statement. For example, depending on the seriousness of the violation, consequences could range from a written warning, to suspension, to employment termination. If workers are found to be selling drugs in the workplace or coercing or intimidating other workers, it may be appropriate to request assistance from law enforcement officials.

There are regions and countries in which written policies are appropriate to the prevailing enterprise culture. In other cases, managers operate their enterprises informally and do not understand the need for a written policy nor see it as a helpful tool for dealing with sensitive issues. This is more likely to be the case not only in smaller enterprises but also in the informal economy.

6.2 Training for supervisors, key staff and workers' representatives

The initial and continuing credibility of a substance abuse policy and programme are highly dependent on the attitudes and actions of supervisors, key staff and workers' representatives. It is essential that these groups receive special training on the rationale and specifics of the programme, provisions of the written policy, signs and symptoms of substance abuse, and basic communication and interviewing skills. Specific training curricula and materials exist or can be developed for this purpose.

Training sessions for supervisors, key staff and workers' representatives should be held regularly, since knowledge and skills, if not used, can fade over time. Training should be presented in several short sessions in order to maintain interest, keep the issue in the forefront and avoid knowledge overload. Also, periodic training will be necessary when new managers, supervisors, staff and workers' representatives are appointed.

Supervisors, professional staff and workers' representatives have distinct roles and responsibilities during the development and implementation of the programme. In addition to the common training discussed above, each group will need specific training relating to its own responsibilities.

Supervisors

The primary tasks of supervisors are to facilitate, monitor and enhance performance, and to intervene in problematic situations. Supervisors need to know what to do should they believe a performance problem or a worker's misconduct is related to substance abuse.

While supervisors should not diagnose, they do need to recognize the signs and symptoms of substance abuse and understand when to refer the employee to a qualified professional for assistance. The information in Annexes I–III gives guidance on this. Supervisors need to know how to confront workers about poor performance or misconduct. By working with a substance abuse professional, supervisors can motivate workers to seek assistance for substance abuse problems in lieu of possible disciplinary action. Supervisors also need training so they can coordinate their actions with other staff and substance abuse professionals while workers are undergoing treatment or during their re-entry process.

Key staff

Key staff members within the organization need specific training related to their responsibilities. For example, occupational health professionals need training in substance abuse and its impact on the workplace and the worker, so that they can initiate appropriate action related to working conditions. Once a troubled employee has been identified, the human resource development staff has a major role in coordinating action between supervisors, workers' representatives, and counselling and treatment professionals. They also need to be familiar with relevant legal provisions and how these affect the actions that management can take. Safety and security officers have an important role in assuring compliance at the worksite level, and need to be trained to recognize signs of substance abuse and the paraphernalia of drug use. The medical staff needs specialized knowledge on dependency and its impact.

If staff members are to provide assessment, counselling, referral and re-entry services, they need professional training: these responsibilities should not be given to untrained staff. It is far preferable to have a substance abuse professional conducting the initial assessment, counselling troubled employees, referring workers to treatment and facilitating re-entry. A substance abuse professional can be hired by the enterprise, or the services of a professional can be obtained from a community or private organization.

Workers' representatives

Since substance abuse programmes deal with the performance and behaviour of workers both within the workplace and, in certain circumstances, outside the workplace, workers' representatives play an important role in implementing prevention programmes. Because of their close relations with workers, they can encourage a positive atmosphere and intervene in problematic situations. Workers' representatives normally have workers' trust and confidence, and are in a better position to identify and intervene with workers and with their families concerning a substance abuse problem. Workers' representatives should know about individual worker rights, especially during assessment, counselling, referral, treatment and re-entry.

6.3 Awareness and education campaigns

For a substance abuse prevention programme to succeed, it is vital that the workers themselves be knowledgeable about substance abuse and committed to the programme's goals. In most cases, this means that comprehensive awareness and education campaigns need to be developed.

Leaflets, brochures, posters, videos and CD-ROM programmes are available commercially. Other material can normally be found within the community or on the Internet. Several organizations have sites that contain information and, in some cases, materials that can be downloaded for use either "as is" or modified. Some of these organizations are listed in Annex VI.

Governments normally have some material available that, while not specific to the workplace, can be very useful in education and awareness programmes. Employers' and workers' organizations may also be a source of information. Some employers or organizations of employers may want awareness and information material adapted for their individual workplaces or specific occupations.

An awareness campaign should be initiated well in advance of the actual implementation of a substance abuse programme. The purpose of the campaign is to inform workers about substances of abuse and their physiological and psychological impacts, the impact of substance abuse on individuals and the workplace, and the availability of counselling and treatment services for those who think they have a problem with alcohol or drugs.

When the prevention programme is initiated, orientation sessions should be held to inform workers of the rationale behind a prevention

> **Box 6.2 Topics that could be included in a worker education programme**
>
> - Laws and regulations covering substance abuse in the workplace
> - The existence of risk and protective factors related to substance abuse and dependency
> - The progressive nature of dependency
> - Communication and interpersonal skills
> - Coping strategies
> - The availability and use of self-help instruments
> - The impact of substance abuse on families
> - Parenting skills
> - Self-help groups
> - Supporting colleagues during and after re-entry into the workplace

programme, the major provisions of the policy statement and the procedures to follow to get help. Orientation information should be made available for new workers.

In most cases, a successful substance abuse prevention programme will depend on changing attitudes and behaviours related to the use and abuse of psychoactive substances. To influence existing attitudes, to build skills and to support behaviour change, an education programme for all workers should be implemented. The programme planning should consider that, depending on the industry sector, occupation and specific working conditions, some categories of workers may have special education needs. Subjects that an education programme might cover are listed in box 6.2.

There are Internet sites that provide information on the availability of self-help groups, professional counselling and treatment. Internet chat rooms and news groups offer the opportunity to share personal experiences and get support in confronting and dealing with substance abuse problems.

Different methods and activities can be used to convey this information, including:

- health promotion activities;
- written material (for example, leaflets delivered in pay envelopes and brochures placed in break areas and cafeterias);

- audio-visual material (for example, a video about the programme on an internal network);
- electronic material (for example, CD-ROMs, interactive programmes on an Internet site, periodic email messages to workers);
- lunch-break discussion sessions;
- company-sponsored competitions and sports activities.

An education and awareness campaign is not a one-time event. Multiple short sessions on various topics using different communication techniques keep the information fresh and provide periodic reinforcement. There are many creative activities which can assist the community in substance abuse prevention. These include mentoring students in the community, sponsoring youth athletic teams, supporting chess or checkers clubs, sponsoring poster contests for the children of workers and holding social events for workers and their families where alcoholic beverages are not served.

6.4 Assistance

Harmful use of alcohol or drugs usually takes a number of years to develop. As it increases, individuals are rarely able to deal successfully with their problems without some level of assistance. The purpose of providing assistance is to interrupt the harmful use of alcohol and drugs, so that workers will make changes resulting in improved health and safety, which in turn improve job performance. Providing workers with access to assistance is often less expensive than incurring the costs of poor performance, lost productivity, disciplinary action, job termination and the recruitment and training of replacement workers.

An initial assessment is done to determine the nature and severity of the problem and the kind of assistance most beneficial to the worker. This usually takes the form of a discussion between a trained professional and the worker. Much harm can result when individuals without professional training attempt to assist, counsel or treat a person with substance abuse problems.

The level of assistance which employers can offer is determined by the services and resources available to them. In some locations, community groups provide professional services. In others, they are provided by private organizations frequently referred to as employee assistance programmes (EAPs). Sometimes, services are only available on a more piecemeal basis. When choosing professional service providers, it is helpful to check the reputation of providers with other enterprises in the community, local or regional chambers of commerce and current or former clients.

Alcohol and drug problems at work

Self-assessment tools exist to help individuals determine whether their alcohol or drug use is problematic or is becoming so. Examples are described in Annex IV. Although Alcoholics Anonymous (AA) is not a treatment programme, it is a self-help programme widely used as a valuable complement to formal assistance initiatives. In many communities around the world, AA or another self-help programme such as Narcotics Anonymous (NA) may be the only resource available for dealing with people who have alcohol or other substance problems.

There are different levels of assistance available to workers, depending on the severity of their substance abuse problems (box 6.3). The financial arrangements regarding assistance can range from employers who provide paid time off work and who pay all assistance costs, to employers who allow workers to use accumulated leave or have time off without pay to pursue assistance at their own expense. Several countries have national healthcare systems which may cover the cost of counselling and treatment.

Workers who are being offered assistance should also be offered confidentiality. Maintaining confidentiality regarding an individual's substance

---

Box 6.3   Assistance for workers with substance abuse problems

- **Counselling** involves discussions between a trained professional and the worker. It is most effective if the substance abuse problem has been identified early, the worker is motivated to change his or her behaviour, and the environment is supportive. Counselling can normally be arranged so that the worker does not miss work or is only away for an hour or two.

- **Referral to treatment** is necessary when the substance abuse problem is more serious and cannot be resolved through counselling. Treatment is more intensive and could require several hours each day or evening on an outpatient basis. Alternatively, when more support and stability are required, the worker may need to participate in an in-patient treatment programme, living in a controlled environment for an extended period of time, usually not less than one month.

- **Rehabilitation** is the process of helping people gain the physical and psychological state and social capabilities they need to deal successfully with daily life. Readjustment, independent functioning and return to work are the final goals of rehabilitation. In general, the ability to obtain and hold a job is regarded as an integral part of this process. Thus, when employers allow workers to undergo treatment without the risk of losing their jobs, rehabilitation becomes much easier.

abuse problem is an important issue in some countries, stemming from the widespread belief that substance abuse is an indication of moral weakness rather than a treatable health problem. Therefore, individuals with substance abuse problems risk being stigmatized and discriminated against, particularly in respect of employment.

6.5 Evaluation

Evaluation should be an integral part of every substance abuse prevention programme. The purpose of conducting an evaluation is to determine whether the programme is accomplishing the goals which were identified during the development of the enterprise policy. A good evaluation will identify both the strengths and weaknesses of the programme and indicate what can be done to improve its operation. The evaluation methodology will dictate what data will be needed and what records should be kept during implementation. It should be determined, therefore, during the programme's planning phase.

There are three levels of evaluation: process evaluation, outcome evaluation and impact evaluation.

Process evaluation

The most frequently used evaluation is process evaluation, which describes what has been done during the implementation phase and measures the results. It could determine, for example:

- the number of managers, supervisors, staff, workers' representatives and workers who have participated in training, education and awareness activities;

- the number of workers who have used the assistance services offered;

- the number of workers who have successfully completed counselling or treatment and have returned to their jobs;

- the number of workers who have been terminated for performance or misconduct associated with substance abuse.

Outcome evaluation

Outcome evaluation determines whether a programme has met its objectives. In an outcome evaluation, baseline data are collected and used to measure the effect of the programme on the issues it was designed to address. An outcome evaluation looks at how the programme has changed the knowledge, attitudes and behaviours of workers who are the recipients of the prevention interventions, and at the aggregate effects of these changes on the enterprise.

The evaluation could examine how the programme has changed the attitudes and intentions of participating workers vis-à-vis drug and alcohol use, and their perception of the risks involved.

At the enterprise level, it could gauge the effect of the programme on the rate of absenteeism, the number of accidents in which drugs and alcohol are involved, healthcare or workers' compensation costs, disciplinary interventions, productivity or any combination of these issues.

Outcome evaluation can be technically difficult and may require resources beyond the means of certain enterprises.

Impact evaluation

This is a more long-term exercise than outcome evaluation. It looks beyond the declared scope and intentions of the programme to see whether there have been consequences which were not planned or expected.

These could, for example, include greater enterprise participation in community efforts to curb the availability of drugs and alcohol, or union initiatives to improve workplace conditions and workers' quality of life.

Evaluation methodologies can range from rigorous to informal, depending on the resources available and the needs of the enterprise. Monitoring the progress of programme implementation should be on-going and can be a regular topic of discussion at steering committee meetings. In this way, adjustments can be made as problems are identified.

Regardless of the kind of evaluation to be performed, provision should be made for obtaining input from managers, supervisors, professional staff, workers' representatives, workers and family members concerning their perceptions of the substance abuse prevention programme and their thoughts on how the programme could be improved.

6.6 Alcohol and drug testing

Though its use has increased since the late 1980s, drug and alcohol testing in the workplace remains a sensitive issue. Much disagreement exists among countries, employers and workers concerning if and how testing should be done. There are also moral and ethical questions that must be taken into consideration in developing and implementing alcohol and drug testing programmes.

On its own, testing for substances of abuse does not constitute a workplace substance abuse programme. Testing, when necessary, should be conducted only as part of a comprehensive programme.

In making a decision whether to have a testing programme, employers, especially in multinational corporations, should be aware that some countries can require testing while others prohibit it. In most cases, the national or local government has not addressed the topic. Additionally, no common international standards and procedures exist. Therefore, it is essential to know the legal requirements of the country in which alcohol and drug testing is being considered.

Testing for alcohol and drugs is not a simple process. Many technical issues must be addressed, such as:

- who will be tested;
- what substances of abuse will be tested for;
- what cut-off levels will be used for each substance;
- when testing will be carried out;
- what will be the frequency of testing;
- what specific tests will be used and procedures followed;
- what actions will be taken in response to a positive test;
- what precautions will be used to protect privacy and confidentiality.

Because of the complexities and sensitivities involved in testing, employers would be well-advised to seek professional advice and assistance in developing testing programmes, to protect both their rights and the right of their workers.

The guiding principles on drug and alcohol testing in the workplace, as adopted by the ILO Interregional Tripartite Experts Meeting on Drug and Alcohol Testing in the Workplace, 10–14 May 1993, Oslo (Hønefoss), Norway, can be found in Annex V of this book.

SUSTAINABILITY STRATEGIES   7

7.1 The importance of sustainability

Perhaps the most difficult task in establishing any new programme, and particularly a substance abuse prevention programme, is developing strategies which guarantee its continued operation. Because each enterprise is unique, strategies will differ from employer to employer.

To be successful, any strategy must ensure the continuing commitment of top management and the availability of resources. These are essential to a prevention programme's long-term sustainability (see box 7.1).

Box 7.1   Key elements of programme sustainability

Maintaining the support of top management

The programme manager and steering committee must establish a mechanism for keeping top management informed. Periodic reporting and briefing sessions can serve as vehicles for presenting the achievements of the prevention programme and its problems, along with recommendations for their solution. Top management can also be invited to participate in awareness and training activities and to take the lead in developing cooperative initiatives within the community and among enterprises.

Ensuring the programme's funding

In some instances, funding for the development and initial implementation of a substance abuse prevention programme may be provided by an organization outside the enterprise (for example, government, a non-governmental organization, an international organization). It is unrealistic, however, to expect external funding to last indefinitely. Employers need to allocate resources for the prevention programme as part of the regular planning and budgetary process of the enterprise.

7.2 Alternative strategies

Choosing an appropriate strategy for programme sustainability will depend on the size and structure of the enterprise, the community, and the industry or sector. Three alternatives are presented below.

Independent programmes

The substance abuse prevention programme can be established as an independent unit that reports directly to top management. This may be the most effective strategy when the nature of the work is so hazardous that any use of alcohol and drugs could be disastrous. In this case, the programme would have a high level of visibility within the enterprise and would command sufficient resources.

Integrated programmes

The substance abuse prevention programme can be integrated into an on-going enterprise programme. For example, the prevention programme could be placed within the human resource management, medical services, welfare, security, legal or occupational safety and health departments. The ILO has found that when prevention programmes are integrated into other programmes they are more durable.

The prevention programme should be integrated into the department best able to further its objective. For instance, this could be human resource management if the objective is improved performance, or medical services if it is to address worker health and rising medical claims.

Making substance abuse prevention part of occupational safety and health (OSH) programmes is frequently the most effective strategy for four reasons:

- National law and regulations frequently mandate OSH programmes at the enterprise level.
- National laws and regulations or negotiated agreements may also mandate that workers' representatives be involved in developing and implementing enterprise programmes, including OSH programmes.
- It can be relatively easy to add substance abuse prevention resources, both financial and human, to OSH programme budgets.
- Occupational health officers can be more easily trained in workplace substance abuse prevention than many other staff members.

Therefore, many of the necessary tools for establishing and implementing substance abuse prevention programmes are already in place in enterprise OSH programmes.

Association of Resource Managers against Drug Abuse (ARMADA)

Establishing an Association of Resource Managers against Drug Abuse (ARMADA) is a strategy that differs significantly from independent or integrated programmes, in that its focus is outside the enterprise.

An ARMADA is a forum of top-level managers from different enterprises who are committed to mobilizing their employees against substance abuse. It is a network for sharing information and experiences, and a mechanism for promoting workplace substance abuse prevention programming in other enterprises.

The ARMADA concept is employer-based, with emphasis on local ownership and company responsibility. Its basic premise is that communication among peers is the most effective means of transmitting information and influencing behaviour.

Establishing an ARMADA within the community requires several separate but related steps.

Identifying the scope of participation

Depending on the size and diversity of the business sector within the community, it may be appropriate to target all enterprises. If this is not practical, consideration could be given to limiting participation, at least initially, to an industry sector (construction, retail, transportation) or to co-located enterprises (industrial park, business district, commercial centre). There is no "right" or "wrong" approach and, if the first configuration doesn't work, the scope can be redefined.

Recruiting employers

There may be several employers within the community who have already implemented substance abuse prevention programmes and who are interested in forming an ARMADA. Employers who do not yet have a prevention programme may also want to join. It is important to make initial contacts at the highest possible organizational level in order to ensure continuing support for the programme.

Developing a structure

An ARMADA can be very informal in structure, with ad hoc meetings hosted by members on a rotating schedule. It can be set up more formally as a working group or a subcommittee of an existing employer group, with a regular meeting schedule and a designated leader. Or, it can be established as a registered non-profit organization or foundation with by-laws, elected leaders and a regular schedule of meetings.

Identifying activities

Members of the ARMADA can focus their activities on the workplace, the family and the community.

Workplace-focused activities can include:
- information and experience sharing;
- worksite visits among members;
- outreach to peers and potentially to other ARMADAs;
- presentations at business meetings and conferences;
- joint programme materials (policies, posters, pamphlets, training curricula);
- joint programme activities (training courses, awareness campaigns).

Family and community activities can include:
- school-based activities, both during and after school hours;
- awareness materials for parents and children;
- training and employment initiatives for recovering workers;
- promotion of laws and regulations which support substance abuse prevention programmes and corollary services within the community;
- recreation facilities and activities.

Providing resources

The level of resources needed to establish and operate an ARMADA is directly related to its structure and functions. For the most basic ARMADAs, the only resources necessary may be members' time and a meeting venue. As an ARMADA becomes more formalized and undertakes a greater number of activities, the need for resources increases. A membership fee may be levied. To keep the need for money at a minimum, members can provide in-kind contributions, including office support, computer equipment, graphic and printing services, and transportation and communications services. If the ARMADA has developed awareness and training materials or training expertise, these can be sold to non-participating enterprises.

There are other sustainability models which can be adopted. There is no single "best practice". The overriding criterion for a successful sustainability strategy is for it to meet the needs of the enterprise.

CONCLUSION 8

8.1 Added benefits

The elaboration of a company policy for a drug- and alcohol-free workplace is a valuable labour–management exercise for any business. The increasingly detrimental effects of alcohol and drug use in the workplace – lost productivity, absenteeism, accidents, compensation claims, lost business opportunities, loss of jobs, loss of skilled workers and strained labour relations – are one problem that workers and employers agree upon as a common dilemma. Approached in this way, the planning and implementation of an effective policy becomes an important form of social dialogue, representing a contract between workers and employers. For this contract to remain valid and for the workplace substance abuse prevention programme to succeed, it must be clear that it applies to everyone in the enterprise from the top down. It must also have the commitment of senior management and the visible support of highly motivated and enthusiastic staff at every level of the company.

Prevention programmes encompassing the whole workforce have been found to have an enormous impact at the enterprise level. They generate a new sense of awareness of substance abuse problems and their consequences for the world of work, providing enterprises with a programme of interventions to address drug and alcohol problems and improve productivity. They influence the philosophy and policy of dealing with drug and alcohol issues in the working environment, and in addition they often give management and workers a new perspective on approaching and managing other workplace problems in partnership. At the national level, they have been a spur to establishing or strengthening policies on drug and alcohol abuse.

The recurring theme of "a programme for all" does not refer only to the internal hierarchy of the company. Substance abuse is a problem which affects communities around the world and one which knows no gender. Although traditionally men have had the greatest problems with substance abuse, women are increasingly using drugs and alcohol and, as primary care-takers, they are directly affected by the abusive patterns of others in their lives, whether parents, siblings, husbands, partners or children. The linkages between the workplace, home and the community inform the structure and goals of a prevention programme, as its positive results can extend through each of these and beyond.

8.2 Some success stories

Programmes in three countries, India, Slovenia and Malaysia, which were supported by the ILO and its partner agencies, illustrate what can be accomplished.

India

Between 1994 and the end of 1999, the ILO implemented a project to develop community-based rehabilitation and workplace prevention programmes in India. Agreements were signed with 12 enterprises, with a combined workforce of over 110,000, and key management officials and their partners in NGOs were all provided with training.

All the enterprises have adopted policies on alcohol and drug use with the involvement of workers' representatives, and are implementing preventive activities, such as the dissemination of their policy to all workers. Other activities include logo and slogan competitions in which workers' children have participated, in-house magazine articles, the distribution of pamphlets and leaflets and the organization of meetings and other events such as plays and shows.

Assistance programmes have also been launched in each of the enterprises in collaboration with the partner NGO. The identification of employees with substance abuse problems is under way and workers have frequent opportunities to consult professional counsellors at the workplace. By the end of the project, one of the participating enterprises had referred over 250 employees for assistance, of whom over 200 had already returned to work.

In these enterprises, for almost the first time in India, addiction is being treated as a health problem. The managers, supervisors, workers' representatives and NGOs involved in workplace programmes all express enthusiasm and expectations are high. Managers from the enterprises feel that the programmes have started to improve the working environment, with employees drinking less and reporting to work on time.

Slovenia

Slovenia was one of six Central and Eastern European countries in which a UNDCP-funded project was carried out, starting in 1995, to mobilize enterprises and workers to prevent substance abuse. Under the guidance of a tripartite advisory board, assistance was provided to six enterprises in Slovenia for the introduction of pilot alcohol and drug abuse prevention programmes.

The participating enterprises have reported that the project has provided an important stimulus for the introduction of a preventive approach in cooperation with the workforce, replacing the former attitude of trying to hide drug and alcohol abuse problems and resorting to dismissal in severe cases. They have also noted a decline in absenteeism, sick leave and accidents, combined with improved productivity and work quality.

Malaysia

Malaysia was one of five countries in which the ILO implemented a Norwegian-funded project in 1998 and 1999 to mobilize small and medium-sized businesses to prevent substance abuse. The primary focus of the project was to prevent the adverse consequences of substance abuse in such businesses, with the aims of enhancing the productivity and competitiveness of the businesses and promoting health and safety in the workplace.

During the course of the project, a total of 44 small and medium-sized businesses implemented the drug-free workplace programme. Workplace programmes were designed to be comprehensive and preventive and to target youth, the group with the highest incidence of drug use, as well as parents, families and employers. The strategies included primary prevention, the rehabilitation of problematic employees and the identification of employees selling drugs. Programmes were implemented as integral components of human resource management functions. Participating enterprises developed a drug-free workplace policy and set up a committee for the implementation of the programme.

In an evaluation survey carried out at the end of the project period, over 80 per cent of respondents reported improvements in business performance, with particular reference to increased productivity, a decrease in work-related accidents, lower absenteeism and a fall in medical costs and compensation. Some 90 per cent of respondents felt that the programme should be sustained and would recommend it to other companies and organizations.

• • •

These success stories illustrate how the advice given here can be applied to great effect in different countries and different employment situations. Using the information provided, an individualized programme can be planned and implemented to work for your company too. In doing so, it is essential to keep in mind that every workplace is different, with special characteristics and resources. The nature of the surrounding community and its interaction with the enterprise and its employees will also have implications for the type of prevention programme to pursue, as will laws and policies at the national level. Experience in a wide range of countries and enterprises has shown, however, that primary prevention is the more effective and less costly way of protecting workers, their families and communities from the impact of substance abuse.

ALCOHOL, ALCOHOLISM AND ALCOHOL ABUSE

ANNEX I

Alcohol

Many people drink alcohol occasionally. Others may drink moderate amounts of alcohol on a more regular basis. For women or those over the age of 65, a moderate amount means no more than one drink per day. For a man, this means no more than two drinks per day. Drinking at these levels usually is not associated with health risks and can help prevent certain forms of heart disease.

Under certain circumstances, however, even moderate drinking is not risk free. At more than moderate levels, drinkers are at risk of serious problems with health, family, friends and co-workers.

What is alcohol?

- A distilled liquid product of fermented fruits, grains or vegetables
- As a constituent of drinks such as wine, beer and spirits, its moderate use is accepted both legally and socially in many cultures around the world
- Also used as a solvent, antiseptic and sedative
- A depressant in small doses which relieves tension and anxiety, and lowers inhibitions
- High potential for abuse

Possible effects

- Reduced motor control
- Sensory alteration
- Anxiety reduction
- Impaired judgement

Symptoms of overdose

- Staggering
- Odour of alcohol on breath
- Loss of coordination
- Slurred speech
- Dilated pupils
- Foetal alcohol syndrome (in babies)
- Nerve and liver damage

Withdrawal symptoms

- Sweating, tremors, insomnia, loss of appetite
- Altered perception, psychosis, fear, auditory hallucinations

Indications of possible misuse

- Confusion, disorientation
- Loss of motor nerve control
- Convulsions, shock, shallow respiration
- Involuntary defecation
- Drowsiness
- Respiratory depression and possible death

Did you know?

- Alcohol is a depressant that decreases the responses of the central nervous system
- Excessive drinking can cause liver damage and psychotic behaviour
- As little as two beers or drinks can impair coordination and thinking
- Alcohol is often used by substance abusers to enhance the effects of other drugs
- Alcohol continues to be the most frequently abused substance among young adults

Annex I

The impact of alcohol

Drinking and driving

Even a very small amount of alcohol can impair the ability to drive safely. Certain driving skills, such as steering a car while responding to changes in traffic, can be impaired by blood alcohol concentrations (BACs) as low as 0.02 per cent. A 72-kilogram (160-pound) man will have a BAC of about 0.04 per cent 1 hour after consuming two beers on an empty stomach.

Interactions with medications

Alcohol interacts negatively with more than 150 medications. For example, taking antihistamines for a cold or allergy and drinking alcohol will increase the drowsiness caused by the medication, making driving or operating machinery even more hazardous.

Interpersonal problems

The more a person drinks, the greater the potential for problems at home, at work, with friends and even with strangers. These problems may include:

- arguments with or estrangement from a spouse and other family members;
- strained relations with co-workers;
- increasingly frequent absence from or lateness to work;
- loss of employment due to decreased productivity;
- committing or being the victim of violence.

Alcohol-related birth defects

Women can prevent alcohol-related birth defects by not drinking alcohol during pregnancy. Alcohol can cause a range of birth defects, the most serious being foetal alcohol syndrome (FAS). Children born with alcohol-related birth defects can have lifelong learning and behaviour problems. Children born with FAS have physical abnormalities, mental impairment and behaviour problems.

Long-term health problems

Some problems, like those mentioned above, can occur after drinking over a relatively short period of time; however, serious health problems can develop more gradually with long-term drinking. The physiological differences between men and women mean that women may develop alcohol-related health problems over a shorter period of time than men, after consuming less alcohol. Alcohol affects many organs in the body and can increase the risk of the following diseases:

- Alcoholic hepatitis, or inflammation of the liver. Symptoms include fever, jaundice (abnormal yellowing of the skin, eyeballs and urine) and abdominal pain. Alcoholic hepatitis can cause death if drinking continues. If drinking stops, this condition is reversible.
- Alcoholic cirrhosis, or scarring of the liver. About 10 to 20 per cent of heavy drinkers develop alcoholic cirrhosis, which can cause death if drinking continues. Cirrhosis is not reversible, but if drinking stops, chances of survival increase considerably, people often feel better and liver function may improve.
- Heart disease, including high blood pressure and some kinds of stroke.
- Cancer, especially cancer of the oesophagus, mouth, throat, and voice box and possibly cancer of the colon and rectum. Women are at slightly increased risk of developing breast cancer if they have two or more drinks per day.
- Pancreatitis, or inflammation of the pancreas. The pancreas helps to regulate the body's blood sugar levels by producing insulin and has a role in digesting the food we eat. Pancreatitis is associated with severe abdominal pain and weight loss. It can be fatal.

Alcoholism and alcohol abuse

For many people, the facts about alcoholism are not clear. What is alcoholism? How does it differ from alcohol abuse?

For most people who drink, alcohol is a pleasant accompaniment to social activities. Moderate alcohol use is not harmful for most adults. Nonetheless, a large number of people get into serious trouble because of their drinking. In many cases the consequences of alcohol misuse are life threatening, and the costs in human terms cannot be calculated.

What is alcoholism?

Alcoholism, also known as "alcohol dependence", is a health problem that includes four symptoms:
- Craving: A strong need or compulsion to drink.
- Loss of control: The inability to limit one's drinking on any given occasion.
- Physical dependence: Withdrawal symptoms such as nausea, sweating, shakiness and anxiety occur when alcohol use is stopped after a period of heavy drinking.
- Tolerance: The need to drink greater amounts of alcohol in order to get drunk. Among heavy, chronic drinkers, the reverse is true: intoxication occurs after just one or two drinks.

Most people do not understand why an alcoholic can't just "use a little willpower" to stop drinking. Alcoholism, however, has little to do with willpower. Alcoholics are in the grip of a powerful "craving" or uncontrollable need for alcohol that overrides their ability to stop drinking. This need can be as strong as the need for food or water.

Although some people are able to recover from alcoholism without help, the majority of alcoholics need assistance. With treatment and support, many individuals are able to stop drinking and rebuild their lives.

Some individuals can use alcohol without problems, but others cannot. One important reason for this has to do with genetics. Scientists have found that having an alcoholic family member makes it more likely that, if you choose to drink, you too may develop alcoholism. Genes, however, are not the whole story. Scientists now believe that certain environmental factors influence whether a person with a genetic risk for alcoholism will develop the health problem. The risk of developing alcoholism can increase, based on how and where a person lives; family, friends and culture; and even how easy it is to get alcohol.

What is alcohol abuse?

Alcohol abuse differs from alcoholism in that it does not include an extremely strong craving for alcohol, loss of control over drinking or physical dependence. Alcohol abuse is defined as a pattern of drinking that results in one or more of the following situations within a 12-month period:

- Failure to fulfil major work, school or home responsibilities;
- Drinking in situations that are physically dangerous, such as while driving a car or operating machinery;
- Having recurring alcohol-related legal problems, such as being arrested for driving under the influence of alcohol or for physically hurting someone while drunk;
- Continued drinking despite having ongoing relationship problems that are caused or worsened by drinking.

PSYCHOACTIVE SUBSTANCES OF ABUSE   ANNEX II

A psychoactive substance is any substance that people take to change the way they feel, think or behave. This includes alcohol, and natural and manufactured drugs. In the past, most drugs were made from plants, such as the coca bush for cocaine, opium poppies for heroin, and cannabis for hashish or marijuana. Now drugs are also produced by synthesizing various chemicals. Drugs can be ingested, inhaled, smoked, injected or snorted.

There is a tendency to present some drugs as less harmful than they actually are, without taking into consideration their long-term consequences and the effects they have on adolescent development, for example, particularly on the development of certain critical functions (cognitive ability and capacity to memorize).

Even occasional use of marijuana affects cognitive development and short-term memory. In addition, the effects of marijuana on perception, reaction and coordination of movements can result in accidents. Ecstasy (MDMA) is often perceived as having little or no negative side effects, but studies show that its use alters, perhaps permanently, certain brain functions, and also damages the liver and other body organs.

Although not regarded as illicit, inhalants are widely abused. Some of these volatile substances, which are present in many products such as glue, paint, gasoline and cleaning fluids, are directly toxic to the liver, kidney or heart. Some produce progressive brain degeneration. Inhalants may cause loss of muscle control; slurred speech; drowsiness or loss of consciousness; excessive secretions from the nose and watery eyes; and damage to the brain and lung cells.

The major problem with psychoactive drugs is that when people take them, they focus on the desired mental and emotional effects and ignore the potentially damaging physical and mental side effects that can occur. **No illicit drug can be considered "safe"**. In one way or another, the use of psychoactive substances alters the normal functioning of the human body and, in the long run, can cause serious damage.

In this annex, drugs are divided into five categories: cannabis, depressants, hallucinogens, narcotics and stimulants.

Cannabis

What is cannabis?

- Cannabis is the name of the hemp plant from which marijuana, hashish and hashish oil are produced. Marijuana is a tobacco-like substance, while hashish consists of resinous secretions of the plant and hashish oil is its extract
- Cannabis extracts are sometimes used in the pharmaceutical industry as compounds of prescription drugs
- Produces a feeling of euphoria followed by relaxation
- Moderate risk of psychological dependence

Possible effects

- Loss of appetite
- Impaired memory, concentration and knowledge retention
- Loss of coordination
- More vivid sense of taste, sight, smell and hearing
- Stronger doses cause fluctuating emotions, fragmentary thoughts, disoriented behaviour and psychosis
- May cause irritation to lungs and respiratory system
- May cause cancer

Symptoms of overdose

- Fatigue, lack of coordination, paranoia and psychosis

Withdrawal syndrome

- Insomnia, hyperactivity, sometimes decreased appetite

Indications of possible misuse

- Animated behaviour and loud talking followed by sleepiness
- Dilated pupils, bloodshot eyes
- Hallucinations
- Distortions in depth and time perception
- Loss of coordination

Marijuana: Did you know?

- Marijuana may cause impaired short-term memory, a shortened attention span and delayed reflexes
- During pregnancy, marijuana may cause birth defects
- Marijuana may cause a fast heart rate and pulse
- Repeated use of marijuana may cause breathing problems
- Marijuana may cause relaxed inhibitions and disoriented behaviour

Depressants

What are depressants?

- Depressants (hypnotics, sedatives and tranquillizers) slow down the central nervous system to suppress neural activity in the brain
- Used medicinally to relieve anxiety, irritability and tension
- Produce state of intoxication similar to that of alcohol (itself a depressant)
- High potential for abuse and development of tolerance
- Combined with alcohol, effects increase and risks multiply

Possible effects

- Sensory alteration, anxiety reduction and intoxication
- Small amounts cause calmness, relaxed muscles
- Larger amounts cause slurred speech, impaired judgement and loss of motor co-ordination
- Very large doses may cause respiratory depression, coma, and death
- Newborn babies of abusers may show dependence, withdrawal symptoms, behavioural problems and birth defects

Symptoms of overdose

- Shallow respiration, clammy skin, dilated pupils
- Weak and rapid pulse, coma, death

Withdrawal syndrome

- Anxiety, insomnia, muscle tremors, loss of appetite
- Abrupt cessation or reduced high dose may cause convulsions, delirium and death

Indications of possible misuse

- Behaviour similar to alcohol intoxication (without odour of alcohol on breath)
- Staggering, stumbling, lack of coordination, slurred speech
- Falling asleep while at work, difficulty concentrating
- Dilated pupils

Selected depressant drugs

- Amylobarbital
- Barbiturates
- Benzodiazepines
- Butobarbital
- Chlordiazepoxide (e.g. Librium)
- Diazepam (e.g. Valium)
- Methaqualone (e.g. Quaalude)
- Nitrous Oxide ("laughing gas")

Hallucinogens

What are hallucinogens?

- Hallucinogens include both naturally occurring and synthetically produced drugs. "Designer drugs" are many times stronger than the naturally occurring drugs they imitate
- Hallucinogens have no known medical use
- Hallucinogens heighten appreciation of sensory experiences, and can produce behaviour changes that are often multiple and dramatic
- Some hallucinogens block sensation to pain, and use may result in self-inflicted injuries

Possible effects

- Rapidly changing feelings, immediately and long after use
- Chronic use may cause persistent problems including depression, violent behaviour, anxiety, and distorted perception of time
- Large doses may cause convulsions, coma, heart and lung failure, and ruptured blood vessels in the brain
- May cause hallucinations, illusions, dizziness, confusion, suspicion, anxiety, and loss of control

- Can have delayed effects: "flashbacks" may occur long after use
- A single use of a designer drug may cause irreversible brain damage

Symptoms of overdose

- Longer, more intense "trip" episodes, psychosis, coma, death

Withdrawal syndrome

- No known withdrawal syndrome

Indications of possible misuse

- Extreme changes in behaviour and mood; a person may sit or recline in a trance-like state or may appear fearful
- Chills, irregular breathing, sweating, trembling hands
- Changes in sense of light, hearing, touch, smell and time
- Increase in blood pressure, heart rate and blood sugar

Selected hallucinogens

- Bufotenine (DMT derivative)
- Indolealkylamines ("magic mushrooms")
- LSD ("acid")
- MDA, MDEA, MMDA
- MDMA ("ecstasy")
- Mescaline
- PCP ("angel dust")

Narcotics

What are narcotics?

- Narcotics are drugs derived from opium, opium derivatives or opium synthetics
- Used medicinally to relieve pain
- Cause relaxation with an immediate "rush"
- High potential for abuse

Possible effects

- Euphoria
- Initial unpleasant effects – restlessness, nausea
- Drowsiness, respiratory depression
- Constricted (pinpoint) pupils

Symptoms of overdose

- Slow, shallow breathing and clammy skin
- Convulsions, coma, possible death

Withdrawal syndrome

- Watery eyes, runny nose, yawning, cramps
- Loss of appetite, irritability, nausea
- Tremors, panic, chills, sweating

Indications of possible misuse

- Scars (tracks) caused by injections
- Constricted (pinpoint) pupils
- Loss of appetite
- Runny nose, watery eyes, cough, nausea
- Lethargy, drowsiness, nodding
- Presence of syringes, bent spoons, needles, etc.

Selected narcotic drugs

- Opium
- Heroin
- Codeine
- Fentanyl

Stimulants

What are stimulants?

- Drugs which speed up the central nervous system to increase neural activity in the brain

- Both synthetically produced and naturally occurring (coca leaves, khat, caffeine and nicotine belong to the latter group), stimulants are used medicinally to treat narcolepsy, obesity and attention deficit disorder
- Used to increase alertness, relieve fatigue, and feel stronger and more decisive, or to counteract the "down" feeling of tranquillizers or alcohol
- High risk of physical and psychological dependence

Possible effects

- Increased heart and respiratory rates, elevated blood pressure, dilated pupils and decreased appetite
- High doses may cause rapid or irregular heartbeat, loss of coordination and collapse
- May cause perspiration, blurred vision, dizziness, a feeling of restlessness, anxiety and delusions

Symptoms of overdose

- Agitation, increase in body temperature, hallucinations, convulsions and possible death

Withdrawal syndrome

- Hunger, apathy, long periods of sleep, irritability, depression and disorientation

Indications of possible misuse

- Excessive activity, talkativeness, irritability and argumentativeness or nervousness

Selected stimulant drugs

- Amfetamines1
- Benzfetamines
- Butyl Nitrite ("poppers")
- Cocaine
- Crack (the smokeable form of cocaine)
- Dextroamfetamine
- Ice (the crystallized, smokeable form of methamfetamine)
- Khat

- Methamfetamine
- Methylphenidate (e.g. Ritalin)
- Phenmetrazine

Cocaine: Did you know?

- A cocaine "high" lasts only about 5 to 20 minutes
- Cocaine use may cause severe mood swings and irritability
- More and more cocaine is needed for each "high"
- Cocaine increases blood pressure and heart rate, which is particularly dangerous in the case of a heart condition
- One use can cause death
- Possession and use are illegal and can result in fines and arrest

Crack: Did you know?

- Crack is almost instantly addictive
- One use could cause a fatal heart attack
- Repeated use may cause insomnia, hallucinations, seizures and paranoia
- The euphoric effects of crack last only a few minutes
- There are more hospitalizations per year resulting from crack and cocaine than any other illicit substances

Ice: Did you know?

- Ice is extremely addictive – sometimes with just one use
- Ice can cause convulsions, heart irregularities, high blood pressure, depression, restlessness, tremors and severe fatigue
- An overdose can cause coma and death
- Stopping use can trigger a deep depression
- Ice causes a very jittery high, along with anxiety, insomnia and sometimes paranoia
- Moderate risk of psychological dependence

SIGNS OF SUBSTANCE ABUSE ANNEX III

A key part of every supervisor's job is to remain alert to changes in worker performance and to assist the worker who is having problems, so that performance improves.

If substance abuse could be contributing to a worker's deteriorating performance, ignoring the situation won't help. Unless some action is taken, the problem will only get worse and could have costly – and possibly disastrous – consequences for everyone.

Workers who abuse alcohol or drugs often hide their problem. Fear and denial are the biggest obstacles in asking for help. Friends, family, and colleagues often begin by playing down the problem. But, as the alcohol or drug problems become more of a burden, irritation and annoyance get the upper hand.

Monitoring performance, not the clinical diagnosis of an alcohol or drug problem, is the job of the supervisor. The supervisor should note and document any deterioration in performance over a period of time. When a *continuing* and *repeated* pattern of abuse begins to appear, "reasonable grounds" for concern are justified and the issues should be addressed based on deteriorating job performance.

Absenteeism	Calling in sick more often, especially for short periods
	Frequent tardiness
	Frequent absence without permission
	Implausible excuses for absence
Absence from work station	Taking extra long breaks
	Taking frequent breaks
	The employee is present, but not at his or her workstation

Loss of concentration	Being physically present, but mentally absent
	Work takes more time and energy than before
	Problems recalling instructions
	Making mistakes on complex transactions
Accidents	Being involved in accidents, or near accidents, more often than other employees
Irregular work patterns	Erratic periods of high and low productivity
	Increasing irresponsibility
	Unpredictable reactions
Externalities	Visibly under the influence
	Smelling of alcohol
	Trembling hands; red swollen face
	Loss of weight and gaunt appearance
	Bloated appearance
	Injection marks on arms
	Worsened personal hygiene
	Seeking opportunities to have a drink
Decreased productivity	Missed deadlines
	More frequent mistakes
	Wasted materials
	Complaints from clients and co-workers
	Poor decisions
Bad relations with colleagues	Reacting strongly to complaints or remarks
	Being moody and suspicious
	Attempting to borrow money
	Evading supervision or control
Other factors	Psychological problems (symptoms of depression, fears, deflection and insomnia)
	Physical problems (stomach complaints, fatigue and memory loss)

SELF-ASSESSMENT TOOLS ANNEX IV

How can I tell if I have a problem with drugs or alcohol?

Drug and alcohol problems can affect every one of us, regardless of age, sex, race, marital status, place of residence, income level or lifestyle. You may have a problem with drugs or alcohol if:

- You can't predict whether or not you will use drugs or get drunk.
- You believe that in order to have fun you need to drink and/or use drugs.
- You turn to alcohol and/or drugs after a confrontation or argument, or to relieve uncomfortable feelings.
- You drink more or use more drugs to get the same effect that you previously got with smaller amounts.
- You drink and/or use drugs alone.
- You remember how last night began, but not how it ended, so you're worried you may have a problem.
- You have trouble at work or in school because of your drinking or drug use.
- You make promises to yourself or others that you'll stop getting drunk or using drugs.
- You feel alone, scared, miserable and depressed.

The CAGE questionnaire

The CAGE questionnaire was developed for early detection of problem drinkers. It consists of only four questions which yield a positive predictive value of over 80 per cent in the description of problem drinkers and self-referred alcoholics in treatment centres. In the workplace situation it obviously does not have the same high predictive value and may not be an effective screening instrument. Nevertheless, it could still be useful as a case-finding instrument in high-risk groups.

The most positive aspect of the CAGE instrument is its simplicity and low cost. Two or more affirmative answers to its four questions are sufficient to identify 80 per cent of the problem drinkers in a high-risk population. The name CAGE is an acronym formed from the initial of the key word in each question. The questions are as follows:

- Have you ever felt that you should Cut down on your drinking?
- Have people Annoyed you by criticizing your drinking?
- Have you ever felt bad or Guilty about your drinking?
- Have you ever had a drink first thing in the morning to steady your nerves or get rid of a hangover (Eye-opener)?

The MAST questionnaire

The Michigan Alcoholism Screening Test (MAST) questionnaire is one of the most widely used tests in the identification or description of problem drinkers in western societies. There is a short version of the MAST called the Brief MAST Questionnaire, which consists of ten questions for self-assessment or interview. The questionnaire is intended for secondary prevention efforts, targeted at groups or individuals who have already been identified as being at risk. The answers to the questions are either "yes" or "no", but the values attributed to the responses vary from 0 to 5. The values are indicated after each possible response and a total score of more than 5 indicates problem drinking. The questions are as follows:

Circle the correct answer:

1. Do you feel you are a normal drinker?   Yes (0)   No (2)
2. Do friends or relatives think you are a normal drinker?   Yes (0)   No (2)
3. Have you ever attended a meeting of Alcoholics Anonymous (AA)?   Yes (5)   No (0)
4. Have you ever lost friends or girlfriends/boyfriends because of drinking?   Yes (2)   No (0)
5. Have you ever got into trouble at work because of drinking?   Yes (2)   No (0)
6. Have you ever neglected your obligations to your family or your work for two or more days in a row because you were drinking?   Yes (2)   No (0)
7. Have you ever had delirium tremens (DTs), severe shaking, heard voices, or seen things that weren't there after heavy drinking?   Yes (2)   No (0)

8. Have you ever gone to anyone for help about your drinking?   Yes (5)   No (0)

9. Have you ever been in hospital because of drinking?   Yes (5)   No (0)

10. Have you ever been arrested for drunk driving or driving after drinking?   Yes (2)   No (0)

WHO alcohol use disorder identification test: AUDIT

In 1987 a World Health Organization (WHO) study developed a valuable instrument for assessing alcohol-related problems on a triple scale: social damage, physical injury, and degree of dependency. The instrument is known as the AUDIT (Alcohol Use Disorder Identification Test), and it seems to be the most promising self-assessment test available for secondary prevention in the context of early identification.

AUDIT is a ten-item self-assessment which can lead people from ignorance of their problem through contemplation to action. It should be made available to all at-risk personnel and to those individuals who are suspected of being problem drinkers. Retesting at three-month intervals can give positive feedback if scores improve. The only danger is that drinkers who score in the caution zone may consider that this means that they do not have a problem.

The Audit Questionnaire

Circle the number that comes closest to the patient's answer:

1. How often do you have a drink containing alcohol?

   Never (0)   Monthly or less (1)   2 to 4 times times a week (2)   2 to 3 times a week (3)   4 or more times a week (4)

2. How many drinks containing alcohol do you have on a typical day when you are drinking?*

   1 or 2 (0)   3 or 4 (1)   5 or 6 (2)   7 to 9 (3)   10 or more (4)

3. How often do you have six or more drinks on one occasion?

   Never (0)   Less than monthly (1)   Monthly (2)   Weekly (3)   Daily or almost daily (4)

4. How often during the last year have you found that you were not able to stop drinking once you had started?

   Never (0)   Less than monthly (1)   Monthly (2)   Weekly (3)   Daily or almost daily (4)

5. **How often during the last year have you failed to do what was normally expected from you because of drinking?**

   Never (0)   Less than monthly (1)   Monthly (2)   Weekly (3)   Daily or almost daily (4)

6. **How often during the last year have you needed a first drink in the morning to get yourself going after a heavy drinking session?**

   Never (0)   Less than monthly (1)   Monthly (2)   Weekly (3)   Daily or almost daily (4)

7. **How often during the last year have you had a feeling of guilt or remorse after drinking?**

   Never (0)   Less than monthly (1)   Monthly (2)   Weekly (3)   Daily or almost daily (4)

8. **How often during the last year have you been unable to remember what happened the night before because you had been drinking?**

   Never (0)   Less than monthly (1)   Monthly (2)   Weekly (3)   Daily or almost daily (4)

9. **Have you or someone else been injured as a result of your drinking?**

   No (0)   Yes, but not in the last year (2)   Yes, during the last year (4)

10. **Has a relative, friend, doctor, or other health worker been concerned about your drinking or suggested you cut down?**

    No (0)   Yes, but not in the last year (2)   Yes, during the last year (4)

* In determining the response categories it has been assumed that one "drink" contains 10 grams of alcohol. In countries where the alcohol content of a standard drink differs by more than 25 per cent from 10 grams, the response should be modified accordingly.

Record the sum of individual item scores here ....

- A score of up to 7 (men) or 6 (women) indicates a safe level of drinking.
- 8–12 (men) or 7–12 (women) indicates a strong likelihood of hazardous and harmful alcohol consumption.
- A score of 13 or more indicates evidence of significant alcohol dependence and further assessment is advised.

DRUG TESTING

ANNEX V

Guiding principles on drug and alcohol testing in the workplace as adopted by the ILO Interregional Tripartite Experts Meeting on Drug and Alcohol Testing in the Workplace, 10–14 May 1993, Oslo (Hønefoss), Norway

Overview

Together, the social partners – employers, workers and their representatives, and governments – should assess the effect of drug and alcohol use in their workplaces. If they conclude that a problem significant enough to require action exists, they should jointly consider the range of appropriate responses in light of the ethical, legal and technical issues enumerated in this document.

A comprehensive policy to reduce the problems associated with alcohol and drug use may cover employee assistance, employee education, supervisory training, information and health promotion initiatives, and drug and alcohol testing. A workplace drug and alcohol testing programme is technically complex and should not be considered without careful examination of all the issues involved.

When a testing programme is being considered, a formal written policy should be developed indicating the purpose for testing, rules, regulations, rights and responsibilities of all the parties concerned. Drug and alcohol testing, as part of a comprehensive programme, should be based on the greatest possible consensus among the parties involved in order to ensure its value.

Background

Reliable analytical methods now exist to detect substances in breath and bodily fluids and tissues. These substances include alcohol as well as other drugs. To ensure programme success, the methods of detection to be used must be of the highest quality and reliability, taking into consideration the purpose of the test. Although the number of competent laboratories is growing, it is recognized that in many countries such facilities do not exist. Policies should therefore be developed to take this into account.

There are two categories of tests: screening and confirmation. The screening test constitutes a rapid but initial stage of the process. However, in the event of a positive test result, confirmatory methods should be used to verify the results. Some legally prescribed drugs may, under certain conditions, be misidentified as illicit substances. Under these conditions, a correct interpretation of the test results is imperative. This highlights the need for high standards not only in technical equipment but also in the training and qualifications of personnel.

It should be recognized that current methods of drug and alcohol testing may involve invasive procedures, which may constitute a risk as well as an intrusion into privacy. In addition, costs for a well-designed drug and alcohol testing programme vary and it is recognized that these may be considerable. It is therefore imperative that these issues be fully considered prior to implementation of any alcohol or drug testing programme.

Assessing the relationship of drug and alcohol use and the workplace

The assessment of problematic use[1] reflected in the workplace should recognize the inherent national, social, cultural, ethnic, religious and gender variables that will affect not only the mode and meaning of use but also the behavioural outcomes evidenced by use. The nature and significance of problematic use must be carefully evaluated.

It is also important to recognize that people often use multiple drugs. Those substances may include, but not be limited to, alcohol, prescribed or over-the-counter medications and illicit or controlled substances.

In addition to the need for a sensitive instrument to assess problematic use of various substances as reflected in workplaces, the *significance* of those defined problems must be assessed. Is it a health, safety, disciplinary or other issue? The identified problematic use should be considered with respect to all of the relevant issues.

Effectiveness of drug testing

The scientific evidence linking the use of alcohol and drugs to negative consequences in the workplace is equivocal. Most evidence, so far, is anecdotal and inferential. Studies are lacking on whether testing programmes reduce possible work difficulties resulting from alcohol and drug use. The available data do not produce sufficient evidence to show that alcohol and drug testing programmes improve productivity and safety in the workplace.

Alcohol and drug testing only recognizes the use of a particular substance. It is not a valid indicator for the social or behavioural actions caused by alcohol and drug use. No adequate tests currently exist which can accurately assess the effect of alcohol and drug use on job performance. There are correlations

---

[1] If use of drugs and alcohol related to the workplace is found after assessment to cause a problem, then for purposes of this document it is considered to be "problematic use". In addressing the use of alcohol and drugs in the workplace, this document does not intend to encourage illegal behaviour.

between behavioural effects and blood alcohol concentrations, but there are variances among individuals. Such correlations have not yet been demonstrated for urine alcohol concentrations, blood drug concentrations or urine drug concentrations.

Programme outcomes

The manner in which organizational, local, national and international drug and alcohol policies and practices are mutually influential is poorly understood. The implementation of a successful policy in one country may have unintended consequences in other countries. Programmes which provide benefits in specific countries may have adverse consequences for other jurisdictions. Since the world is becoming more closely linked by the existence of multinational corporations and international trade agreements, for example, countries and enterprises should examine more closely the international impact of their initiatives. Drug and alcohol policies must be individualized to meet the needs of particular users. One locality's policy cannot meet all users' needs.

Concern about the consequences of drug and alcohol use in the workplace should be addressed in a comprehensive strategy. If testing is considered as one of the elements in a comprehensive strategy, the intended outcomes as well as unintended effects should be considered. Review of these effects can assist in the decision to include testing in a programme strategy or not. If testing is to be included, this review can assist in determining the nature and extent of the testing to be carried out.

Some of the intended outcomes may include:

- Assistance in the development of a comprehensive programme to improve safety and security as well as to reduce potential legal liabilities.
- A comprehensive productivity and quality assurance programme including reduction of absenteeism.

Some of the unintended outcomes may include:

- Deterioration of the work environment through fear, mistrust, polarization between management and workers, lack of openness, and increased social control.
- Not following legal and ethical rules.
- Breaches of confidentiality.
- Adverse effects on individuals as a result of errors in testing.
- Decrease in security of employment.

Legal and ethical issues

There are ethical issues of fundamental importance in determining whether to test for drugs or alcohol. Is testing warranted? If so, under what circumstances?

Recognizing that the situation differs in each country and each workplace, ethical issues are one of the most important concerns to be resolved before any testing is undertaken. Rights of workers to privacy and confidentiality, autonomy, fairness and the integrity of their bodies must be respected, in harmony with national and international laws and jurisprudence, norms and values. Workers who refuse to be tested should not be presumed to be drug or alcohol users.

The need for testing should be evaluated with regard to the nature of the jobs involved. With some jobs, the privacy issue may be determined to outweigh the need to test.

As a protection to workers, positive test results should be subject to independent medical review. For those workers whose positive test results reflect problematic drug or alcohol use, participation in a counselling, treatment or self-management programme should be encouraged and supported.

Various national laws, customs or practices may require that employees who test positive are referred for treatment, assigned to other work or that other means to ensure their security of employment are used.

Specific procedures should be developed which demonstrate a programme's capacity to comply with existing national laws and regulations. Such regulations may include:

- legislation on workplace drug and alcohol testing,
- labour law,
- medical confidentiality laws.

Drug and alcohol testing must be placed within the larger context of the moral and ethical issues of collective rights of society and enterprises, and of individual rights, as embodied in the Universal Declaration of Human Rights and international labour standards.

Other rights are also important: examples here are the right to choose one's own doctor, the right to representation if needed, the right to notification that testing will be carried out as part of a pre-employment screening programme, and the right to information on test results.

It is assumed that the participants in any work situation have rights and responsibilities which may have been agreed upon. Drug and alcohol testing programmes should fit within existing arrangements for ensuring the quality of work life, workers' rights, the safety and security of the workplace, and employers' rights and responsibilities (e.g. protection of the public interest).

Programme organizers should be sensitive to the potential for any adverse consequences of testing (e.g. harassment, unwarranted invasion of privacy). Workers should have the right to make informed decisions about whether or not to comply with requests for testing.

Safeguards should be installed to eliminate any potentially discriminating impact from testing. The testing programme should be conducted in a non-discriminatory manner in compliance with the appropriate legislation and regulations. In those jurisdictions with a constitutional right to work, efforts should be made to enable the person to remain in the workforce.

Annex V

Programme organization and development

Setting up a programme

Where problems in workplace performance exist, a number of responses may be considered. If the problems are related to the effects of alcohol and drug use, the balance of the corrective strategy should lean towards education and prevention. Partners in the workplace must consider whether employee assistance programmes are available within the enterprise, through the trade union or external associations in the community at large.

Assistance programmes should be voluntary, "broad brush" approaches which are capable of addressing a wide range of health promotion issues. If drug testing is an option within the assistance programme, a number of methods exist including pre-employment, post-accident, reasonable suspicion, post-treatment, random or voluntary testing.

In any case, drug testing should be viewed only as part of a systematic approach which includes assessment, information on the effects of various levels of substance use, education concerning the elements of a healthy lifestyle and a programme of reintegration into the workplace for problematic drug or alcohol users.

The response selected must be directly related to the workplace problems to be addressed. The objectives to be met by the testing programme must be clearly defined and articulated. Before drug testing is selected as an appropriate response, there must be clear evidence that testing can reasonably be expected to achieve its intended goals.

In this context, it is especially important to determine that the technical capacity for state-of-the-art testing procedures exists and is used. Analysis of test results must take into account the differences between alcohol and other drugs.

Programme policy statement

The written policy should detail the procedures to be adopted by the testing programme and these should be agreed upon by all the social partners. The policy should clearly identify the purpose of testing and the uses for the results. It should indicate any laws or regulations concerning drug and alcohol testing that may apply. If needed, a summary statement could explain how the programme intends to comply with those laws or regulations.

It should emphasize workers' rights, employers' rights, public rights and individual rights. It should identify the substances to be tested for and how these substances will be detected. It should describe the testing method and the relevance of that method to the results. It should explain the laboratory procedures and the analytical methods used by the laboratory. It should detail how the testing programme is to be organized, the level of administrative support required, the technical expertise needed, who will carry out the tests and with what equipment.

Any changes to the policy, because of new conditions or because other substances are being tested for, should only take place with the agreement of all the social partners.

Administrative structure

The testing programme's administrative structure, areas of responsibility and lines of authority should be clearly delineated in written form and should be made freely available. A specific organization officer should have primary responsibility for the programme's operations. The manner in which the testing programme fits within the organization's larger administrative structure should be clearly stated. The qualifications of programme personnel should also be clearly stated.

Administrative procedures should be established to ensure that procedures have been followed correctly. These procedures should address the status of the tested individual and the responsibility of the organization during the time that test results are being analysed.

Confidentiality

Standards to protect the privacy of the workers and to ensure the confidentiality of test results should be specified. Among these, the following guidelines should be observed:

1. The identity of the worker should be kept confidential.
2. The records concerning the worker should be kept in a secure location.
3. Separate authorization by the worker should be obtained before the release of each test result, specifying the tested substances.
4. Signed authorization to divulge information about a worker to third parties should name the specific individual(s) who will receive the information.
5. A separate authorization should be obtained for each intended recipient of information about the worker.
6. Authorization forms should be witnessed.
7. Policies concerning the confidentiality of the testing programme should be communicated to relevant parties.

Programme linkages

A mechanism should be in place for communicating test results to the tested person. Appropriate mechanisms should be established to allow that person to be referred for assistance when indicated and when the person consents.

Policy options/purposes

The purposes of any drug and alcohol testing programme should be specified in writing. Among the most common purposes given for testing programmes are:

- investigations of accidents and incidents;
- referral for assistance;
- deterrence;
- meeting legal and regulatory requirements;
- communicating an organization's policy.

The form of drug and alcohol testing in a particular programme should be explicitly tied to the purposes of the programme. For example, many forms of testing may be adopted to meet regulatory requirements. Although there may be some disagreement regarding the value and utility of any particular form of testing, it seems that:

- Reasonable suspicion and post-accident testing are most clearly linked to investigative purposes.
- Pre-employment, post-treatment monitoring and voluntary testing may be most appropriate if the organization wishes to refer persons who have been identified as drug and alcohol users for assessment and consultation.
- Pre-employment, random, transfer, promotional and routine scheduled testing may be compatible with deterrence purposes.

It is imperative to establish written criteria governing when to apply one of the options listed above. In addition, the frequency and duration of testing assigned as part of assistance monitoring and/or a return to work programme should be prescribed in the overall testing policy. When a pre-employment test is utilized, it may be part of a comprehensive medical examination used to determine fitness for work.

Determining which drugs to test for

Decision-making on testing should be flexible, and existing conclusions should be reviewed periodically. Decisions to test for alcohol and drugs should be made only when reliable and valid initial and confirmatory testing services or facilities are available. These facilities should protect confidentiality and, for forensic purposes, ensure the chain of custody. All positive results should be confirmed prior to notification or any other action.

Several criteria should be considered in selecting which substances to test for:

- the prevalence rates and the consequences of use in the workplace;
- the likelihood of harm to health due to use of various substances;
- the likelihood of substance use affecting work-related behaviour.

Programme evaluation and review

Ongoing evaluation and review are essential to ensure that a testing programme is able to attain the objectives for which it was established. The plan for monitoring and evaluation should be set out when a testing programme is designed.

The evaluation plan should:

- be based upon the written goals and objectives established for the programme;
- identify means to determine whether or not the programme is being implemented as intended; and
- establish criteria and mechanisms for determining the impact and effect of the testing programme.

Evaluation plans should adhere to acceptable standards. Results of evaluations should be made available to all relevant parties.

Technical and scientific issues

In many parts of the world no programmes exist for the accreditation of testing laboratories. In order to ensure the highest accuracy and reliability of the testing programme, standard operating procedures should be in place to document the manner in which specimens are handled, instruments are checked for proper functioning and quality control is carried out. Accuracy and reliability must be assessed in the context of the total laboratory system. If the laboratory uses well-trained professional personnel who follow acceptable procedures, then the accuracy of results should be very high.

The working group recognizes that national and international standards are lacking. It recommends that the ILO consider using such standards as those developed by the National Institute on Drug Abuse and the College of American Pathologists as a basis for developing international standards.

Extreme caution must be exercised in the testing procedures. Testing specimens beyond the authorized list of drugs for other types of analysis (e.g. HIV, other disease criteria or pregnancy) should be expressly prohibited. Additionally, the possible impact of a positive result on an individual's livelihood or rights, together with the possibility of a legal challenge of the results, should set this type of testing apart from most clinical laboratory analysis. All workplace alcohol and drug testing should be considered as a special application of analytical forensic toxicology. That is, in addition to the application of appropriate analytical techniques, the specimens must be treated as evidence and all aspects of the testing procedure should be documented and available for examination.

The purpose for which testing is conducted will often dictate the specimen of choice. Typically blood is examined when impairment issues are addressed, while urine is examined when drug use is being questioned. In many countries the law may require consent prior to submitting to the sampling procedure. Before any sample is collected the employee should be informed as to the collection procedure, the drugs that will be tested for, the associated medical risk and the use of results. Provisions should be made for the protection of the personnel responsible for specimen collection.

At present, urine appears to be the best specimen for analysis in the context of detecting drug use in the workplace. Specimen collection procedures should be done in such a way that the privacy and confidentiality of the donor is protected as well as the integrity of the specimen.

Blood can be used to detect the presence of alcohol and most drugs. However, the invasiveness and discomfort of the sampling procedure, the requirements for a trained phlebotomist and provision for emergency medical assistance make blood a less desirable specimen for workplace testing.

In terms of testing for alcohol, the breath is the most commonly used specimen. Equipment is readily available and breath can be collected in a non-invasive manner.

At this time, insufficient data exist to support a recommendation for an alternative specimen such as hair, sweat or saliva.

Initial screening and confirmation methods must be based on different principles of analytical chemistry or different chromatographic separations.

Quality assurance and quality control protocols should be in place before the initiation of the analytical procedures. These procedures should encompass all aspects of the testing process, from specimen collection through reporting of the results to final disposition of the specimen. Quality assurance procedures should be designed, implemented and reviewed to monitor each step of the process.

A positive result does not automatically identify an individual as a drug user. The results should be reviewed, verified and interpreted by a medical expert. Prior to making a final decision, the medical expert would check all medical records, examine other medical explanations for a positive test result, and conduct a medical interview with the individual (including the individual's medical history). This would determine whether a confirmed positive result could be explained by the use of legally prescribed medication.

Before making the final interpretation of the test result, the individual should be given the opportunity to discuss the test results with the medical reviewer. If there is a legitimate medical explanation for the positive test, the result should be reported as negative and no further action should be taken.

Recommendations for action and research

The expert working group recommends that:

- Research should be undertaken to evaluate the relationship between use of drugs and alcohol and job safety and productivity.
- Research should be initiated to evaluate the costs and benefits of testing programmes. Evaluations should be done to study the costs and benefits for each of the parties, including social, economic and psychological costs and benefits.
- The ILO should consider the need for developing international standards for drug and alcohol testing and laboratory certification.

WEBSITES ANNEX VI

Canadian Centre on Substance Abuse (CCSA)
The CCSA promotes informed debate on substance abuse issues and encourages public participation in reducing the harm associated with drug abuse. It disseminates information on the nature, extent and consequences of substance abuse, and supports and assists organizations involved in substance abuse treatment, prevention and educational programming. The site is available in English or French. http://www.ccsa.ca/

Center for Substance Abuse Prevention (CSAP)
CSAP's mission is to make the connection between substance abuse prevention, research and practice. It is part of the Substance Abuse and Mental Health Services Administration (SAMHSA), the agency of the United States government charged with improving the quality and availability of prevention, treatment and rehabilitative services in order to reduce illness, death, disability and cost to society resulting from substance abuse and mental illnesses. The site is available in English or French. http://www.samhsa.gov/centers/csap/csap.html

European Monitoring Centre for Drugs and Drug Addiction (EMCDDA)
The mission of the EMCDDA is to provide the European Community and its Member States with objective, reliable and comparable information at European level concerning drugs and drug addiction and their consequences. This site is available in each of the European Community languages, and is linked to the EMCDDA's online database of information on European drugs-related legislation, ELDD. http://www.emcdda.org/home.shtml

International Labour Organization (ILO)
The ILO's Workers' Health Promotion and Well-being at Work programmes are designed to help meet the ILO's commitment to further among the nations of the world those programmes which will achieve adequate protection for the

life and health of workers in all occupations. Drugs and alcohol comprise one section of this site and contain information on the ILO's programme. http://mirror/public/english/protection/safework/drug/index.htm

ILO Substance Abuse in the Workplace Database
The ILO Substance Abuse in the Workplace Database provides key information on issues related to alcohol and drugs in the workplace. The documents indexed cover topics related to programme planning, policy development, legal issues, statistics, and research. Both English and French documents are included. Abstracts are written in the language of the document. http://www.ccsa.ca/ilo/

National Clearinghouse for Alcohol and Drug Information (NCADI)
The National Clearinghouse for Alcohol and Drug Information is the information service of the Center for Substance Abuse Prevention of the US Department of Health & Human Services. NCADI is the world's largest resource for current information and materials concerning substance abuse. http://www.health.org/

National Drug and Alcohol Research Centre (NDARC)
The Centre, established at the University of New South Wales in 1986, primarily aims to increase the effectiveness of treatment for drug and alcohol problems in Australia. http://www.med.unsw.edu.au/ndarc/

National Institute of Alcohol Abuse and Alcoholism (NIAAA)
The NIAAA supports and conducts biomedical and behavioural research on the causes, consequences, treatment and prevention of alcoholism and alcohol-related problems. NIAAA also provides leadership in the national effort to reduce the severe and often fatal consequences of these problems. NIAAA is one of 18 institutes that comprise the National Institutes of Health (NIH), the principal biomedical research agency of the United States government. http://www.niaaa.nih.gov/

National Institute of Drug Abuse (NIDA)
NIDA's mission is to lead the United States in bringing the power of science to bear on drug abuse and addiction. This charge has two critical components: the first is the strategic support and conduct of research across a broad range of disciplines; the second is to ensure the rapid and effective dissemination and use of the results of that research to significantly improve drug abuse and addiction prevention, treatment and policy. NIDA is part of the National Institutes of Health. http://www.nida.nih.gov/

Rutgers University Center of Alcohol Studies Database
The Alcohol Studies Database contains citations of over 55,000 documents indexed by the Rutgers University Center of Alcohol Studies since 1987. The primary focus of the database is on research and professional materials dealing

with alcohol, its use and related consequences. A growing amount of literature on other drug use/abuse has been added in recent years. The database also includes a small collection of educational and prevention materials, including audiovisuals, suitable for schools, parents, community workers and the general public. http://www.scc.rutgers.edu/alcohol_studies/ (key phrase: "workplace and alcohol")

South African National Council on Alcoholism and Drug Dependence (SANCA)
The SANCA Resource Centre offers access to a library that specializes in addictions. It contains a wide range of information on the topics of drug and alcohol addictions, employee assistance programmes and other related issues. http://www.sn.apc.org/sanca/sirc.htm

United Nations International Drug Control Program (UNDCP)
UNDCP is responsible for concerted international action for drug abuse control. Its website provides a comprehensive list of publications in various languages on issues surrounding narcotic drugs and psychotropic substances. http://www.odccp.org/

US Department of Labor Working Partners Initiative
Working Partners for an Alcohol- and Drug-Free Workplace is a one-stop source for information about workplace substance abuse prevention. The Working Partners programme provides employers with introductory resources and tools to address the problematic use by employees of any substance – including but not limited to alcohol, illegal drugs, and prescription and over-the-counter drugs. http://www.dol.gov/asp/programs/drugs/about.htm

Virtual Clearinghouse on Alcohol, Tobacco and Other Drugs
The Virtual Clearinghouse on Alcohol, Tobacco and Other Drugs is a global partnership to facilitate access to information on policy, research and compiled statistics. The emphasis is on access to the "fugitive" or "grey" literature, such as national drug strategies, that are generally not included in existing indexing and abstracting services. The Documents Database is a collection of links to full-text documents in their original language, but these can be read with a linked automatic translation service. http://www.atod.org/

World Health Organization (WHO)
The online catalogue of the WHO describes over 30 of the agency's publications, most of which are available in various languages, covering many aspects of drug and alcohol use and abuse. http://www.who.int/dsa/cat98/subs8.htm

PRINT AND AUDIO-VISUAL RESOURCES

ANNEX VII

1. Print resources

General information

Allsop, S.; Phillips, M.; Calogero, C. (eds.). 2001. *Drugs at work* (Melbourne, IP Communications).

Ames, G.; Grube, J.; Moore, R. 1997. "The relationship of drinking and hangovers to workplace problems: An empirical study", in *Journal of Studies on Alcohol*, Vol. 58, No. 1, pp. 37–47.

Blum, T.C.; Martin, J.K.; Roman, P.M. 1993. "Alcohol consumption and work performance", in *Journal of Studies on Alcohol*, No. 54, pp. 61–70.

Brecht, J.; Poldrugo, F.; Schaedlich, P. 1996. "Alcoholism: The cost of illness in the Federal Republic of Germany", in *PharmacoEconomics*, No. 10, pp. 484–493.

Butler, B. 1993. *Alcohol and drugs in the workplace* (Toronto, Butterworths).

Coambs, R.; McAndrews, M. 1994. "The effects of psychoactive substances on workplace performance", in S. Macdonald and P. Roman: *Research advances in alcohol and drug problems, Volume II. Drug testing in the workplace* (New York, Plenum Press), pp. 77–102.

Dietze, K. 1992. *Alkohol und Arbeit* (Zurich, Orell Füssli Verlag).

English, H. et al. 1995. *The quantification of drug-caused morbidity and mortality in Australia, 1992* (Canberra, Commonwealth Department of Human Services and Health).

Guppy, A. 1993. *Workplace systems for managing drug-related problems* (Bedfordshire, Cranfield Institute of Technology).

Heller, D.; Robinson, A. 1993. *Substance abuse in the workforce: A guide to managing substance abuse problems in the workplace* (Ottawa, Canadian Centre on Substance Abuse).

Holcom, M.; Lehman, W.; Simpson, D. 1993. "Employee accidents: Influences of personal characteristics, job characteristics and substance use", in *Journal of Safety Research*, Vol. 24, pp. 205–221.

International Labour Office. 1992. *Drugs and alcohol in the maritime industry*, Report of the ILO Interregional Meeting of Experts, Sep.–Oct. (Geneva).
—. 1995. *Management of alcohol- and drug-related issues in the workplace*, code of practice (Geneva).
Leigh, J.P. 1996. "Alcohol abuse and job hazards", in *Journal of Safety Research*, No. 1, pp. 17–52.
Mullahy, J.; Sindelar, J. 1996. "Employment, unemployment and problem drinking", in *Journal of Health Economics*, No. 15, pp. 409–434.
National Health and Medical Research Council (NHMRC). 1997. *Workplace injury and alcohol* (Canberra).
Normand, J. 1993. *Drug use by the workforce: Magnitude, detection, impact and intervention*, paper presented to the Comité permanente de lutte aux drogues, Montreal, May.
Normand, J.; Lempert, R.; O'Brien, C. (eds.). 1994. *Under the influence: Drugs and the American workforce* (Washington, DC, National Academy Press).
Osterberg, E. 1992. "Current approaches to limit alcohol abuse and the negative consequences of use. A comparative overview of available options and an assessment of proven effectiveness", in *The negative social consequences of alcohol use* (Norwegian Ministry of Health and Social Affairs, Oslo, Aug.), pp. 266–299.
Shahandeh, B. 1985. "Drug and alcohol abuse in the workplace: Consequences and countermeasures", in *International Labour Review*, Vol. 124, No. 2, pp. 207–223.
Shain, M. 1990/91. "My work makes me sick: Evidence and health promotion implications" in *Health Promotion*, winter, pp. 11–12.
Smith, J.P. 1993. *Alcohol and drugs in the workplace: Attitudes, policies and programmes in the European Community*, Report of the ILO in collaboration with the Health and Safety Directorate, Commission of the European Communities (Geneva).
Webb, G. et al. 1994. "The relationship between high-risk and problem drinking and the occurrence of work injuries and related absences", in *Journal of Studies on Alcohol*, No. 55, pp. 434–446.
Wilsnack, R.W.; Wilsnack, S.C. 1992. "Women, work, and alcohol: Failures of simple theories", in *Alcoholism: Clinical and Experimental Research*, Vol. 16, No. 2, pp. 172–179.

Prevention programmes

Cook, R.F.; Youngblood, A. 1990. "Preventing substance abuse as an integral part of worksite health promotion", in *Occupational Medicine: State of the Art Reviews*, Vol. 5, No. 4, Oct.–Dec., pp. 725–738.
Di Martino, V.; Gold, D.; Schaap, A. 2002. *SOLVE: Managing emerging health-related problems at work* (Geneva, ILO).
International Labour Office. 1992. *Manual on the design, implementation and management of alcohol and drug programmes at the workplace* (Geneva).

—; United Nations International Drug Control Programme. 2001. *Drug and alcohol abuse prevention programmes in the maritime industry* (Geneva).

Schneider, R.; Colan, N.; Googins, B. 1990. "Supervisor training in employee assistance programs: Current practices and future directions", in *Employee Assistance Quarterly*, Vol. 6, No. 2, pp. 41–55.

Testing

Macdonald, S.; Roman, P. (eds.). 1994. *Drug testing in the workplace: Research advances in alcohol and drug problems*, Vol. 11 (New York, Plenum Press).

Ontario Human Rights Commission. 1990. *Policy statement on drugs and alcohol testing* (Toronto, Nov.).

UNDCP. 1993. "Drug testing in the workplace", in *Bulletin on Narcotics*, Vol. XLV, No. 2.

Assistance programmes

Addiction Research Foundation. *Guidelines to creating an employee assistance program* (Toronto).

Blum, T.C.; Martin, J.K.; Roman, P.M. 1992. "A research note on EAP prevalence, components and utilization", in *Journal of Employee Assistance Research*, Vol. 1, No. 1, pp. 209–229.

Kronson, M.E. 1991. "Substance abuse coverage provided by employer medical plans", in *Monthly Labor Review*, Vol. 114, No. 4, Apr., pp. 3–10.

Ritson, B. 1990. "Services available to deal with problems faced and created by alcohol users", in Norwegian Ministry of Health and Social Affairs: *The negative social consequences of alcohol use* (Oslo, Aug.), pp. 227–259.

US Department of Labor. 1991. *What works: Workplaces without alcohol and other drugs* (Washington, DC, Oct.).

Legal issues

Daintith, T.; Baldwin, R. 1993. *Alcohol and drugs in the workplace: A review of laws and regulations in the Member States of the European Community* (London, Institute of Advanced Legal Studies).

Denenberg, T.S.; Denenberg, R.V. 1991. *Alcohol and other drugs: Issues in arbitration* (Washington, DC, Bureau of National Affairs).

Industrial Relations Service. 1992. "Drink and drugs in the workplace", in *Industrial Relations Legal Information Bulletin*, No. 460, Nov., pp. 2–9.

Trubow, G. 1991. *Privacy law and practice* (New York, Matthew Bender), pp. 9–18 to 9–21.

2. Audio-visuals

CNS Production. 1992. *Drugs and work performance*, video, 27 minutes (Ashland, OR).

SAMPLE POLICIES ON SUBSTANCE ABUSE

ANNEX VIII

Sample policy 1

Reproduced from the ILO publication *Drugs and alcohol in the maritime industry*.

ALL VESSELS UNDER THE CONTROL OF …LINE/COMPANY

Policy objectives

(a) To prevent drug abuse amongst all employees as part of the …Line/Company's commitment to the health and welfare of its employees, operational safety and environment.

(b) To provide an education programme for employees on the dangers and consequences of drug abuse.

1) Eligibility

This policy covers all shore-based employees and seafarers employed by …Line/Company.

2) General background

…Line/Company regards the health and safety of customers, and employees ashore and afloat to be of paramount importance and is totally committed to the highest safety standards.

The use of psychoactive drugs can jeopardize personal safety, the safety of others, working areas and environment. Contrary to widely-held belief, there are no safe drugs. All can lead to mental and physical deterioration, affect behaviour adversely and impair the ability to work.

Abuse of any drugs can lead to addiction, overdose and death. Intravenous drug abuse is well recognized as a source of transmission of hepatitis B and HIV/AIDS.

It is essential to have a firm Drug Policy to ensure the well-being and safety of all persons ashore and afloat including our customers and therefore the highest standards are required by all staff.

The company acknowledges that the Merchant Shipping (Health and safety: General Duties, XX) Regulations (1984 XX) place general health and safety duties and responsibilities on both employees and employers.

The presence of illicit drugs onboard ship can have serious legal implications for companies and individuals including, fines, imprisonment and the detention of ships.

3) Procedure

3.1 The ...Line/Company will provide an education programme for employees on the dangers and consequences of drug abuse. This will include the distribution of leaflets, posters, and the encouragement of onboard discussions in welfare and safety committee meetings.

3.2 Employees should be aware that any employee *will be dismissed* from their employment if it is established that they have been involved in any of the following, either onboard ship or elsewhere:

   i) Found to be involved in the trafficking of drugs.

   ii) Found to be in possession of drugs or used drugs other than prescribed by Medical practitioner. Employees who have been prescribed drugs must inform management when appointed to a ship and advise the Master upon joining.

   iii) Convicted in a court of law of any offence connected with drugs or drug abuse. It should be clearly understood that any employee will be dismissed, upon conviction, whether they are on or off a Crew Agreement, on board or elsewhere, wherever the offence occurs.

3.3 Accordingly, as an employee of ...Line/Company, if it comes to the attention of management that any of the above circumstances apply, the appropriate disciplinary procedures will be instituted.

   i) If the facts relate to conduct onboard, the relevant procedures will be set out in the appropriate Disciplinary Procedures. In this regard, only the Master has the authority to discipline or discharge an employee from the ship.

   ii) If the facts relate to conduct ashore, or while the employees are off the Crew Agreement, there will be a disciplinary hearing held ashore with the right to appeal to a higher level of shore management.

3.4 As an indication of commitment to this Policy ...Line/Company will;

   i) Reserve the right to carry out pre-employment drug screening.

ii) Require any employee suspected of drug abuse to undertake a drug test. A system of random drugs screening carried out by a reputable, independent company will take place where the Master determines that there is a "reasonable cause" to carry out such testing, or where an individual involved in a drug-related incident requests a drug test.

Any drug screening procedure will include a referral to a medical review officer in the event of a positive result, at no cost to the employee.

3.5 The contract clearly stipulates that continuation of employment is based upon meeting statutory medical standards which includes a clear history of no drug abuse. Deliberate failure or refusal to participate in a drug test as required in this policy will be considered to be a refusal to accede to a company medical examination and may result in disciplinary action, including dismissal.

3.6 The …Line/Company may, where an individual with drug-related problems voluntarily declares that they have a drug-related problem, assist the individual in seeking help to overcome such problems with counselling, treatment and rehabilitation, In the event the individual responds positively to any prescribed counselling, treatment and rehabilitation, the …Line/Company will look favourable to re-employment of the individual.

By enforcing this policy, the …Line/Company is acting in accordance with the agreed approach of the Industry generally, including both Seafarers' and Shipowners' organizations.

If any employee has any queries regarding this policy, please discuss them in the first instance with the appropriate Port Personnel or Training Manager.

REMEMBER

PERSONS UNDER THE INFLUENCE OF DRUGS ARE A HAZARD TO PASSENGERS, THEIR SHIPMATES, TO THEMSELVES, THE FREIGHT, CUSTOMERS AND THEIR SHIP.

ANY EMPLOYEE IS LIABLE TO DISMISSAL FOR POSSESSING OR USING DRUGS OR TRAFFICKING IN DRUGS EITHER ONBOARD OR ANYWHERE ASHORE.

DRUG TAKING PUTS THE USER'S HEALTH, AND POSSIBLY LIFE, AT RISK AND CAN LEAD TO ADDICTION.

Sample policy 2

Reproduced with the kind permission of MICO Bosch, Bangalore, India.

Policy statement

We, the management, Union and Employees of Motor Industries Company Limited, Naganathapura Plant, strongly believe that alcohol and drug abuse is a major hazard. We therefore will work towards preventing alcohol and drug abuse, **providing early assistance** to needful employees and commit ourselves to **helping needful employees in helping themselves** in becoming healthy and productive members of our prestigious organization.

Need for the policy

Alcohol/drug abuse is not only a health hazard but also has serious consequences on productivity at the workplace and family at large. If this is not attended to, it will lead to addictions resulting in physical suffering, financial loss and mental agony for the employees and their families, and have an adverse impact on others in the Company. Hence the need to have a prevention programme to **educate** the workforce about the serious consequences of alcohol and drug abuse.

Objectives of the policy

1) To prevent alcohol/drug abuse and its ill effects
2) To educate employees on the consequences of alcohol/drug abuse
3) To enable employees to become better members of family and society
4) To assist employees to enable them to overcome their addiction
5) Early introduction of necessary therapy
6) Restoration of health
7) To enable the employee to retain the job.

Scope of the policy

This policy will apply to all employees across all levels/departments/offices of the Company. This policy covers alcohol- and drug-related problems which impair work performance, attendance, conduct, reliability, safety and health.

Responsibilities

- To provide information to employees, superiors, peers and family members on matters relating to alcohol and drug abuse through necessary interventions.

Annex VIII

- To provide sufficient training to employees, superiors and peers in matters relating to assistance and early intervention.
- To motivate those affected to avail early intervention, assistance, treatment and after care by the Company. In this regard, the Company's medical services will coordinate with the Personnel Department to provide necessary support.

Regulations and control

The employees shall be aware that the intake of alcohol and drugs have detrimental effects on their self-control, judgment and ability. It may also have adverse effects on their attendance and productivity, making them liable for suitable disciplinary action as per the Company's certified Standing Orders and other rules and regulations.

No employee shall report for duty under the influence of alcohol, under any circumstances. The consumption of alcohol within the office or work premises is strictly prohibited.

Guidelines to handle alcohol- and drug-related medical problems

- Employees who feel that they have alcohol- or drug-related problems are encouraged to seek help in their own interest from the Company without fear of discrimination.
- Any employee who seeks help will get necessary support.
- Employees, who have a responsibility to safeguard their colleagues' interests, will be equipped to assist them.
- All managers and Union representatives will be assisted to perform their roles in implementing the established procedures.
- All employees are expected to follow the policy.

Implementation

The working committee shall be constituted, which will establish programmes and procedures; periodically review and evaluate the programme and make suitable changes wherever necessary by involving all concerned.

Company support

1) Company help shall consist of two time-staggered offers, conditions and actions.
2) If the superior suspects that one of his/her employees is at risk of addiction or is addicted, he/she holds a confidential discussion with the employee in question, after consulting the welfare service, the works' medical service or

an assistant. He/she shall then inform the welfare service, the works' medical service or the assistant, of the outcome of these discussions. These services and assistants shall then advise the employee and, if necessary, indicate to him/her ways of counteracting his/her addiction or risk of addiction.

Three months after the first discussion with the employee in question, the superior shall inform the personnel department of the employee's suspected addiction or risk of addiction if, after consulting the welfare service, the works' medical service or an assistant, he/she is of the opinion that the employee in question is not prepared to take suitable action in order to counteract his/her addiction or risk of addiction.

3) The personnel department shall inform the union about the employee's addiction or risk of addiction in good time before further discussions are held. The personnel department shall summon the superior, the works' doctor, a representative of the welfare service and, if the employee so wishes, the member of the union responsible for welfare matters, to attend the first discussion. The employee may be advised to consult an addiction advisory, to join a self-help group or to undergo outpatient or inpatient treatment. None of the discussions, in accordance with paragraphs 2 and 3, Section 1, shall be taken as a reason to implement disciplinary action or unilateral personnel action in individual cases.

4) If the employee in question has not taken the required action recommended to him/her in the preceding discussion, and if the addiction or risk of addiction has effects on duties (such as high rate of absenteeism, poor or reduced performance, or other violations of employment-contract obligations), in accordance with the contract of employment, the first employment-contract consequence may be drawn after the second discussion (warning or written request to desist). A condition is then imposed upon the employee in question that he/she must take the recommended help or assistance. The works' security shall be informed of the special risk relating to the employee.

5) If the employee in question still does not meet his/her obligations from the Contract of Employment and if he/she does not take the recommended action, the second disciplinary measure may be implemented after expiry of a period of a further four weeks (warning or written request to desist). It must be pointed out to the employee in question that it is intended to dismiss him/her if the offer to help him/her is not accepted and if obligations from the Contract of Employment are still not met.

6) If the employee in question does not meet the conditions and if he/she is still not able to perform his/her duties resulting from the Contract of Employment properly, further disciplinary action will be taken.

7) If notice of dismissal is issued, a pledge must be made to the employee in question that he/she will be reinstated should he/she prove, within a period of one year subsequent to leaving the Company by means of a doctor's

certificate, that the withdrawal treatment has been completed successfully and that the employee in question can be considered as totally abstinent on the date upon which the Company is to reinstate him/her.

The promise of reinstatement applies only to a claim to employment in an equivalent job or place of work, but not necessarily in the previous job or place of work. In the event of reinstatement, the time spent working with the Company shall be credited to the employee, after a period of two years, provided the employee affected has not suffered a relapse up to this point.

Procedure in case of relapse

1) In the case of relapse after successful withdrawal treatment or other help or assistance, a decision shall always be taken, individually allowing for all circumstances of the individual case, as to what action is to be taken.

    In general, the action shall commence with a discussion in accordance with S 3, para. 3 of this Company Agreement. The periods stipulated in the Company Agreement maybe shortened if advocated by the welfare advisory and the works doctor.

2) In cases in which reinstated employees suffer relapses during the first two years subsequent to reinstatement, the Contract of Employment may be terminated, allowing for the rights of co-determination of the Works Council, without requiring further discussions or action in accordance with S 3 of this Company Agreement beforehand. A pledge shall be made to the employee in question to the effect that a labour court settlement shall be offered if a dismissal protection suit is filed. In this settlement, a pledge shall be made to the employee to:

    i) Reinstate him/her, initially limited to a period of two years, after successful withdrawal treatment and to terminate this time-limited Contract of Employment, allowing for the rights of co-determination of the Works Council, only if the employee suffers a further relapse. In turn, one further pre-condition for reinstatement is that the employee proves, by way of a doctor's certificate, within a period of one year subsequent to his/her leaving the Company, that the withdrawal treatment has been completed successfully and that he/she can be considered as totally abstinent on the date of scheduled reinstatement by the Company.

    ii) Grant an indefinite Contract of Employment, after expiry of the deadline, if the time-limited Contract of Employment has not yet been terminated by the appointed date. The previous time with the Company shall be credited on the date of granting of an unlimited Contract of Employment.

3) The legal, collectively-bargained Company provisions shall apply to dismissals for other reasons.

Sample policy 3

Reproduced with the kind permission of Cititel Hotel, Penang, Malaysia.

Objectives

- To maintain a safe, healthy, productive and dadah-free[1] work environment.
- Covers the use and abuse of dangerous drugs, prescription drugs and other forms of substance abuse, including alcohol.

General provisions

- The Hotel recognizes that use and abuse will impair the ability to perform properly and will have serious adverse effects on safety, efficiency and productivity and the good social environment.
- The Hotel strictly forbids the use, possession, sales or trafficking in drugs.
- No employee shall be under the influence of drugs whilst on duty or carry drugs on the Company work locations.
- The Hotel reserves the right to refuse the entry of "Prescribed Drugs", if the use of such drugs will affect the safety and efficiency of the work environment.
- An employee arrested and charged by the authorities for being under the influence of, possession, sales or trafficking in drugs within on outside Company work locations will be subjected to disciplinary action which may result in dismissal.

Drug search

The Hotel reserves the right to search any employee and/or his personal belongings within the Company work locations. Assistance of the Police or other enforcement agencies may be called upon.

Rehabilitation procedure

The Hotel recognizes drug dependency as a treatable condition. Employees with drug dependency problems should voluntarily request for rehabilitation assistance from the Company. This includes leave without pay for not more than 14 days for detoxification, if necessary. Employees returning from rehabilitation will be required to participate in a Company-approved after-care programme.

---

[1] "Dadah" is the usual Malaysian term for drugs.

Annex VIII

- The employee who refuses the rehabilitation assistance from the company or the employee who is under the rehabilitation assistance but does not follow through is subject to disciplinary action including dismissal from employment.
- Each employee is allowed to request for rehabilitation assistance from the Company only once. The employee who is found with a recurrence of drug dependency problem is subject to disciplinary action which may include stern warning, downgrade or dismissal from employment.

Drug tests

- All employees and sub-contracted personnel on Company work locations will be subjected to random urine tests for drugs, irrespective of position.
- Test for Cause.
- Test after accidents at work locations.
- Candidates considered for employment will have to go through a drug test.

Infringement of policy

Infringement of any policy provisions, employee subject to disciplinary action including dismissal from employment.

Work locations

Includes all company premises; Company's and clients' vehicles, staff hostel.

Sample policy 4

Reproduced with the kind permission of Akzo Nobel.

AKZO NOBEL CORPORATE DIRECTIVE 13.5

Occupational Health and Safety Management System

Purpose

The purpose of this directive is to ensure the implementation of an Occupational Health and Safety Management System and maintaining the established system.

Definition

The Policy Statement on Health, Safety and Environment includes statements on Health and Safety. Based on these statements the following definition for an *Occupational Health and Safety Management System* has been derived:

A management tool to ensure that Akzo Nobel's Occupational Health and Safety policy is effectively and efficiently implemented by:

A. setting auditable objectives
B. defining authorities and responsibilities
C. providing high competence and knowledge
D. collecting, continuously evaluating, and safeguarding data required for long-term health surveillance of our employees in each department necessary.

Directive

An Occupational Health and Safety Management System is mandatory for all locations. A Substance Abuse Policy shall be an element of the Occupational Health and Safety Management System. Implementation shall be finalized no later than the year 2002.

Responsibilities

The general managers of business units and service units are responsible for the implementation of this directive within their organization.

References

- Akzo Nobel Policy Statement on Health, Safety and the Environment

Annex VIII

- Occupational Health and Safety Management System Standard (OHSMS) like:
  - BS 8800: 1996
  - Care System Master (Akzo Nobel Chemicals SHERA 1996)
  - Det Norske Veritas
  - OHSAS 18001 issued by BSI (1999)
- Akzo Nobel Health, Safety and Environmental Manual, Fourth Edition, January 2002
- Akzo Nobel Guidelines for an Alcohol and Substance Abuse Policy

AKZO NOBEL GUIDELINES FOR AN ALCOHOL AND SUBSTANCE ABUSE POLICY

1. Summary

The Akzo Nobel Policy Statement on Health Safety and the Environment defines Occupational Health and Safety as "the prevention of injuries and harm to the health of our employees and other persons as a result of our activities". This statement is the basis for the Alcohol, Drugs and Medicines Policy of Akzo Nobel.

The aim of these guidelines is to provide a framework for all locations to establish or update their local policies on ADM-abuse.

The main principles of the ADM-policy of Akzo Nobel are given below: on the basis of these principles all Akzo Nobel locations have a local responsibility to develop and implement their own ADM-policy.

The local ADM-policy is considered part of the local Occupational Health and Safety Management system which is mandatory for all locations no later than the year 2000. As such the ADM-policies will be subject to HSE-audits (see Corporate Directive 13.2).

Principles: The following principles serve as a guide for the development and implementation of the local ADM-policies:

1. The general managers of business units and service units are responsible for the development, implementation and enforcement of the local ADM-policy.
2. Akzo Nobel considers the ADM-policy as a part of the Occupational Health and Safety Management System: the policy aims at improving safety at work and work-related activities (e.g. transport) and at the reduction of absenteeism, sickness, disability, accidents and incidents, productivity decreases and/or poor performance as a result of ADM-use.
3. The local ADM-policy needs to be adapted to national law, to local and national customs and cultures and to the hazards and risks related to the

business. A uniform approach within the Akzo Nobel organization is not likely to work, but the application of common principles ensures consistency in Occupational Health and Safety practice.

4. Important elements of any ADM-policy are:
   - clear rules regarding the use, sale or possession of alcohol, drugs or illegal medicines at work;
   - information and education on the effects of ADM-abuse on occupational health and safety;
   - training of line-managers to recognize and discuss performance problems with employees at an early stage;
   - a programme for assistance, treatment and rehabilitation (Employee Assistance Programme = EAP);
   - disciplinary measures if treatment and rehabilitation fails.

5. It is recommended to involve occupational health professionals and human resource managers in the development of a local ADM-policy and to consult workers' representatives (e.g. labour council) prior to implementation.

6. It is recommended that contractors are required to have a similar policy in place in order to qualify for work.

7. In some countries screening and testing procedures (e.g. pre-employment testing, random testing for use of alcohol and drugs, testing after accidents) may be required to reduce legal exposure. Testing procedures involve moral, ethical and legal issues of fundamental importance. Testing procedures should be validated by a Medical Review Officer in order to ensure a legally valid interpretation of test results.

8. Testing procedures may jeopardize open communication between employer and employees on alcohol, drugs and medicine abuse. In case local custom is permissive towards testing procedures, it is recommended to initiate testing only after full implementation of training, education, treatment and rehabilitation. In such case testing is used as a control measure to ensure compliance with a well established ADM-policy, (Further info. on testing procedures can be provided by Central Office–Corporate Health).

2. Introduction

The ADM-policy will be most effective when it is considered by all parties involved as an essential element of the Occupational Health and Safety Policy; full implementation should be considered a joint effort between employer and employees. The employer should provide and maintain a safe and healthy workplace and take appropriate actions, including the adoption of an ADM-policy. The employees should cooperate with the employer to prevent accidents at work due to abuse of alcohol, drugs or medicines.

The objective of the ADM-policy is:
- to safeguard the health and safety of all workers;
- to prevent accidents;
- to improve productivity and efficiency in the company;
- to assist employees who are experiencing problems related to the use of alcohol, drugs or illegal medicines.

It is well recognized that ADM-problems may be related to illnesses like depression or to personal distress, e.g. caused by marital problems or family difficulties, on-the-job or off-the-job stress, financial problems, legal difficulties etc. Therefore the Akzo Nobel ADM-policy is based on the provision of counselling, treatment and rehabilitation to employees.

A guiding principle for the ADM-policies within the Akzo Nobel organization is to prevent "a blame culture" and to stimulate an open culture where performance problems can be discussed: it is evident that the use of alcohol, drugs and (illegal) medicines becomes a problem for the organization when work performance deteriorates. It is a key element that early signs of performance deterioration are discussed between employer and employee. If performance does not improve within a set period, there will be repercussions. This method of "constructive confrontation" is a critical success-factor of any Akzo Nobel ADM-policy: on one hand the company offers assistance to employees to solve their problems, on the other hand the employee is confronted with the facts of deteriorated performance and the possible consequences if performance does not improve.

For many managers and supervisors it is not difficult to recognize deterioration of performance, however they find it difficult to discuss and to document ADM-problems and deterioration of performance with employees. By stimulating an open culture, companies will lower the threshold to discuss and deal with ADM-problems specifically and performance problems in general. The earlier ADM-problems are recognized, the better prognosis for full recovery and rehabilitation.

Further information

Further information and assistance for the development and implementation of the ADM-policy may be obtained from the Corporate Health Adviser and/or the local Occupational Health Physician.

3. Development of an Alcohol, Drugs and Medicines Policy for the workplace

3.1 Joint effort of social partners

The ADM-policy of the locations should preferably be developed as a joint effort between management and representatives of the employees (e.g. labour council). Where feasible also health professionals and/or human resource professionals with knowledge of ADM-problems should be involved in the development of the policy.

3.2 Contents of the ADM-policy

A policy for the management of alcohol, drugs and (illegal) medicines in the workplace should include:

- measures to prohibit or restrict the availability of alcohol, drugs or (illegal) medicines in the workplace;
- prevention of ADM-problems in the workplace through information, education, training and any other relevant programmes;
- identification and referral of employees with ADM-problems;
- a procedure for intervention, treatment and rehabilitation of employees with ADM-problems;
- a procedure for disciplinary measures if the ADM-policy is violated.

3.3 Assessment of the effects of ADM-use in the workplace

It is useful when management and representatives of the employees jointly assess the effects of ADM-problems in the workplace. The following indicators could provide useful information for identifying and assessing the size of the problem:

- national and local surveys on ADM-consumption rates;
- surveys that were carried out in similar companies;
- absenteeism in terms of unauthorized leave and lateness;
- use of sick leave;
- accident rates;
- personnel turnover;
- alcohol consumption on company grounds;
- opinions and experiences of supervisors, employees, health professionals and safety advisors.

Although the above indicators can only provide a rough estimate on the extent of ADM-problems within the company-location, the data will be useful to clarify the needs and priorities in the set-up of ADM prevention and assistance programmes.

4. Prevention through information, education and training

Information, education and training programmes about alcohol, drugs and (illegal) medicines should preferably be integrated into broad based health programmes. Information, education and training programmes should be considered the first step of the introduction of a new (or changed) ADM-policy.

This programme usually includes:

a) For all employees

- information on the effects which drugs and alcohol may have on health and fitness for duty and the related risks to safety and the environment;
- information on the services available to assist employees with alcohol- and drug-related problems, including information concerning referral, counselling, treatment and rehabilitation;
- information on the company's ADM-policy including the clear rules at the workplace.

b) For supervisors and managers

- to identify and discuss impaired performance and behaviour which may indicate that the services of an employee assistance programme (EAP) or health professional might be useful and to give information on these services to the employee;
- to support the needs of a recovering employee and to monitor performance when the employee returns to work.

5. Clear rules at the workplace

The ADM-policy should contain clear rules regarding (a) fitness for duty and (b) restrictions on alcohol, drugs and (illegal) medicines in the workplace.

a) Fitness for duty

- the employee is responsible for being "fit for duty". Employees should be prohibited from being on company-locations with impaired performance due to alcohol, drugs or (illegal) medicines.

b) Restrictions

1. Alcohol

- Management of the locations should consider – after consultation with (representatives of) employees – restricting or prohibiting the possession, consumption and sale of alcohol in the workplace, including in the canteen and dining area and during business meetings (lunch and dinner).
- Management should apply the same restrictions or prohibitions with respect to alcohol to all levels of employees, so that there is a clear and unambiguous policy.
- As a general rule alcohol should not be used during working hours; exceptions should be clearly defined.
- Management should not formally or informally support behavior which encourages or facilitates the harmful use of alcohol or the use of drugs in the working environment.

2. Drugs
- The use, possession, distribution or sale of drugs or illegal medicines should be strictly prohibited.

3. Medicines
- When the employee uses legally prescribed medicines which may result in significant impairment, the employee should consult a qualified occupational health professional. This professional should then determine fitness for duty with any necessary restrictions.

4. It is well recognized that the restrictions or prohibitions referred to above may vary significantly depending on the risk to safety and the environment and the national, cultural and social traditions.

5. Unannounced searches for alcohol, drugs and/or illegal medicines on company locations, particularly in safety and environmentally sensitive positions, could serve as a suitable first line deterrent for possession.

6. Identification

Line-managers and colleagues have an important role to play in recognizing ADM-problems and in convincing employees to seek treatment.

The identification of individual employees with ADM-problems may be conducted at three levels:

a) self-assessment by the employee, facilitated by information, education and training programmes;

b) informal identification by friends, family members or colleagues, who suggest that the worker who appears to have a problem should seek medical advise and to follow appropriate treatment accordingly;

c) formal identification by the employer/line-manager on the basis of actual facts of impaired work performance.

It can be expected that shortly after the implementation of the ADM-policy, the majority of employees with ADM-problems will be identified via formal identification. However by creating an open culture where ADM-problems can be discussed and where assistance is provided and where a "blame culture" is avoided, the threshold to discuss ADM-problems with health professionals or line-managers will be lowered, resulting in earlier informal identification (b) or self-assessment (a).

7. Employee assistance, treatment and rehabilitation

a) Health character of ADM-problems

The same considerations should be given to employees with ADM-problems as to those having an illness, with the objective to minimize risks to work-performance, safety, health and environment.

b) Job security

Employees who seek treatment and rehabilitation should not place their employment in jeopardy by doing so. Company benefits which apply in the case of illness should be available.

After recovery of the ADM-problems they should enjoy normal job security and opportunity for transfer and promotion. Exception to the principle of job security may be justified when the Occupational Health Physician determines that the employee is no longer fit for a given job (e.g. for safety reasons), in such cases alternative work may be considered.

c) Assistance to employees

Coordination of assistance to employees who have ADM-problems will vary according to the size and nature of the location, as well as according to national law, health-care and social security systems.

Such an Employee Assistance Programme (EAP) can be established by the employer or (preferably) as a joint programme between employer and employee representatives.

In small Akzo Nobel locations the EAP may be organized as a point of initial assessment and referral to care-givers in the community, such as medical doctors, specialists in alcohol and drug-counselling, treatment and rehabilitation, or community based organizations.

In some larger Akzo Nobel locations appropriate professionals may be available to offer confidential assistance to employees – to help them with their ADM-problems and a whole range of other problems liable to cause personal distress, including marital or family difficulties, depression, anxiety or stress, financial problems or legal difficulties. Counselling, treatment and rehabilitation programmes should be adapted to the individual needs of the person concerned.

In the case of ADM-abuse it is recommended that a treatment contract is signed between employee and employer prior to referral to the EAP. If the employee does not live by the agreements made in the treatment contract, the occupational health department will inform the employer that the EAP will be discontinued. In such case the employer may take disciplinary action.

d) Reintegration

After formal treatment a recovery programme may include an ongoing period of after-care, which can be a crucial **part of the EAP.**

When a qualified health professional determines that an employee is successfully pursuing treatment or has successfully completed treatment, the employee should be offered a transfer or retraining opportunity when the return to his current job is not appropriate.

e) Privacy and confidentiality

All records and information regarding ADM-problems should be treated in the same way as other confidential health data. Occupational health personnel should be permitted to communicate to the employer whether an employee is fit for work, fit with restrictions, or not fit for work, and the duration of the disabling health condition.

8. Disciplinary measures

"Constructive confrontation" is a key element to assist employees with malperformance due to ADM problems: on one hand the company offers the EAP, on the other hand – in such case – a treatment contract will be signed between employer en employee.

If a worker fails to cooperate with the treatment contract or if work-performance does not improve after intensive counselling, treatment and rehabilitation, the employer may take disciplinary action.

In accordance with national law and practice, disciplinary measures concerning ADM should be elaborated by the employer, preferably in consultation with employees and/or their representatives. Such measures should be communicated to the employees so that they clearly understand what is prohibited and the sanctions for violation of the rules.

Dismissal should normally occur in the following circumstances, however such action is subject to local employment law:

- failure to comply with the rehabilitation procedures
- the use, possession, distribution or sale of illegal drugs on company locations
- the use of alcohol on company locations, unless previously authorized

9. Contractors

Contractors should be obliged – preferably by means of the contract – to comply with the ADM-policy of the location. The Akzo Nobel companies should ensure that the ADM-policy of contractors is compatible with the company's policy and that the policy is observed by contractor personnel while working on company grounds.

The company may require from contractors that they enforce in their contracts with subcontractors that they maintain an ADM-policy which is at least equivalent to the company's policy.

APPENDIX A Model for an Alcohol, Drugs and Medicines Policy

(A policy statement should outline the programme to be developed in the company. It should take into account the social and cultural norms, the nature of the businesses and the accessibility to professional resources etc. The presented model should therefore be adapted and modified to fit with the local circumstances.)

The Alcohol Drugs and Medicines Policy of the company is part of the Occupational Health and Safety Management System, which is aimed at prevention of injuries and harm to the health and safety of our employees and other persons as a result of our activities. Moreover the company is committed to maintain productivity and efficiency through a healthy workforce. This is considered a joint effort between employer and employees.

Annex VIII

For the company it is imperative that employer and employees are acutely aware of the adverse effects that the use of alcohol, drugs and (illegal) medicines may have on health, safety and well-being of employees and their colleagues, including a deterioration of work-performance. Through the implementation of the ADM-policy the company wants to minimize the risks of ADM-use. Therefore the following policy is an integral element of the terms and conditions of employment.

- The ADM-policy applies to all employees of the company; contractors and subcontractors are required to maintain an ADM-policy which is equivalent to the company's policy.

- The company does consider dependence on ADM as a treatable condition: through information, training and education, efforts are made to ensure prevention, early identification, treatment of ADM-problems, thus facilitating a good prognosis. Employees who seek treatment and rehabilitation will not place their employment in jeopardy. After recovery of the ADM-problems they will enjoy normal job security and promotion. The company benefits which are applicable in the case of illness will be available.

- Through the Employee Assistance Programme employees will be assisted to receive treatment and support during rehabilitation. All medical records and information regarding ADM-problems will be treated as all other confidential health data.

- Referrals to the Employee Assistance Programme can be made on the basis of self-referral by the employee or by colleagues or supervisors who consider that problems are due to ADM. An employee is allowed to refuse referral by supervisors, however in such case the employee denies himself the opportunity of treatment and rehabilitation which may lead to disciplinary measures in the case of impaired performance.

- If a worker fails to cooperate with the treatment programme or if work-performance does not improve after finishing the Employee Assistance Programme, the company may take disciplinary action.

- Being at work whilst impaired by alcohol, drugs and/or illegal medicines is strictly prohibited. The use, possession, distribution or sale of drugs and illegal medicines on company grounds is strictly prohibited, also the use of alcohol on company grounds is prohibited.

- If the employee uses legally prescribed medicines, which may result in significant impairment, this individual should consult the company health adviser who will determine fitness for work with or without restrictions for work.

- The security department may conduct unannounced searches for alcohol, drugs or illegal medicines on company locations.

- This ADM-policy is part of the Occupational Health and Safety Management System of the company.

APPENDIX B Definitions

Abuse	1. Being on company locations while being impaired by alcohol, drugs or illegal medicines. 2. Use of alcohol, drugs or illegal medicines on company locations where ADM-use is prohibited by company rules.
ADM-problems	The term alcohol, drugs and medicines problems can be applied to any social, medical, personal and work problems that result from alcohol consumption or drug/medicines taking.
Dismissal	Termination of employment.
Employer	The legal "body" who employs workers.
Employee Assistance Programme	A professional and confidential programme for referral, counselling and treatment of employees for a broad spectrum of personal problems.
Employee Representatives	Persons who are recognized as such by national law or practice, e.g. labour council.
Fit for work	A good mental and physical condition which enables employees to perform their duties in a safe and productive way.
Impaired performance	Any factual and objective signs that an employee is not able to perform his duties in a safe and efficient way due to any loss or abnormality of a psychological or physical function.
Illegal medicines	Medicines that have not been prescribed or dispensed in the name of the user.
Legal medicines	Prescribed drugs that have been dispensed in the name of the user.
Occupational Health Services	Health services which have essentially a preventive function and which are responsible for advising the employer, as well as the employees and their representatives, on the requirements for maintaining a safe and healthy working environment.
Treatment contract	If an employee is referred to the Employee Assistance Programme by his or her supervisor because of impaired performance due to ADM- abuse, the terms and conditions of the EAP may be stated in a treatment contract (Appendix D).

Annex VIII

APPENDIX C Graphic display of 'CONSTRUCTIVE CONFRONTATION'

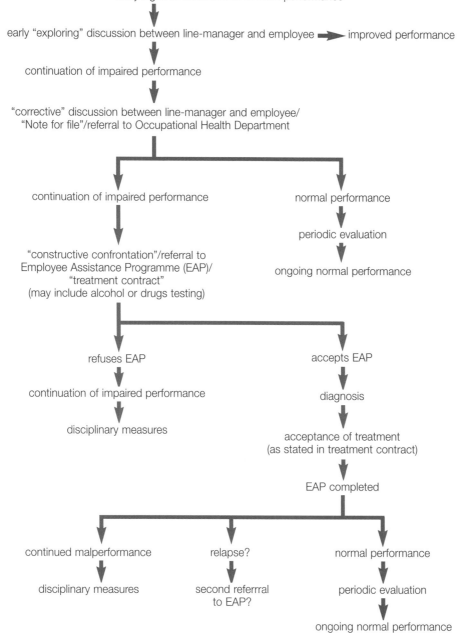

APPENDIX D Model for a "Treatment Contract"

To Mr/Mrs....   Confidential

As discussed earlier with you (see "Note for file" dated dd/mm/yy), your performance has deteriorated over the past (period of time). Apparently this is also caused by the use of alcohol (or drugs). This would be sufficient reason for termination of employment, however our company is prepared to offer you a final opportunity to improve your performance.

1.   From dd/mm/yy onward you will be declared unfit for duty (e.g.) as a driver; your supervisor will arrange restricted duties for you.

2.   You are strongly advised to consult our Occupational Health Department for referral to the Employee Assistance Programme. After a period of 6 months we expect you to be fully fit for duty again.

3.   The Occupational Health Department will be (co-)supervisor of the course of the treatment programme, in order to monitor the course of actions.

4.   If you do not comply during these 6 months with the treatment contract and if the company is convinced that there is no reasonable chance that you will resume your own duties, the company will initiate the termination of your employment.

5.   If it appears that – after the treatment period of 6 months – you did not cease alcohol (or drugs) abuse, or if it becomes clear to the company that you are not sufficiently fit for your own duties, the company will also initiate the termination of your employment.

6.   If there is a recurrence of alcohol (or drug) abuse after you have resumed your duties, your employment will be terminated immediately.

7.   After you have resumed your duties you will agree on and cooperate with a performance review with your supervisor at two-monthly intervals

We request you to sign this letter for agreement . We express our sincere hope that the treatment will be successful in order to enable you to continue your duties in a satisfactory way.

Signature: Manager/supervisor

Signature: Employee

Annex VIII

APPENDIX E Suggestions for implementation

ADM-programmes in the workplace need to be carefully introduced, supported by information and education.

It is recommended to ask support from a consultant for the provision of education and training and to advise on the implementation-plan for the ADM-policy. The non-profit organization ALCON (Stichting Alcohol Consultancy Netherlands) could serve as a qualified consultant for a multinational organization.

The following elements may be included:

1. The development of an ADM-policy should preferably be a joint effort of employer and employee representatives: discussion, acceptance and agreement is of vital importance for the successful introduction of an ADM-policy.
2. Integration of information, education and training programmes about ADM into other activities.
3. Use of posters, hand-outs, booklets concerning the company's ADM-policy.
4. Training of supervisors how to recognize, document and discuss early signs of deterioration of work performance – **this is a critical success factor!**
5. Work-shops on management of ADM-problems for health and human resource professionals involved in the Employee Assistance Programme.
6. Provide a "help-desk function" at Corporate Level where companies may discuss specific problems during development and implementation of the ADM-policy. This "help-desk" could contribute to consistency and exchange of "best practices" of ADM-policies within AKZO Nobel companies.

References

- Akzo Nobel Guidelines for an Alcohol, Drugs and Medicine Policy (1998)
- Health, Safety and Environment Audits (Corporate Directive 13.2)
- *Management of alcohol- and drug-related issues in the workplace* (ILO, Geneva, 1996).